# Elephant Destiny

*Martin Meredith*

# Elephant Destiny

*Biography of an Endangered
Species in Africa*

PublicAffairs

*New York*

Originally published as *Africa's Elephant* in Great Britain in 2001
by Hodder and Stoughton.
A division of Hodder Headline.
Published in the United States by PublicAffairs™,
a member of the Perseus Books Group.
All rights reserved.
Printed in the United States of America.

Meredith, Martin.
Elephant Destiny : a biography of an endangered
species in Africa/ by Martin Meredith.
p. cm.
Includes bibliographical references (p.).
ISBN 1–58648–077–4
1. African elephant. I. Title
QL737.P98 M473 2003
599.67'4—dc21
2002036776

First PublicAffairs Edition 2003
2  4  6  8  10  9  7  5  3  1

For Tilly, Daisy, Ellie and Susie

# Contents

# Contents

# Acknowledgements

The broad nature of this book has meant that I have relied on the work of many other authors. Starting with Aristotle, the father of elephant science, in the fourth century BC, the trail leads through a landscape of Roman historians, medieval naturalists, Victorian poets, colonial hunters and game wardens, culminating in the endeavours of modern biologists. In recent times, the discoveries of field scientists such as John Perry, Irven Buss, Richard Laws, Iain Douglas-Hamilton, Cynthia Moss, Joyce Poole and Katy Payne have transformed our understanding of the elephant world. I am especially indebted to Iain and Oria Douglas-Hamilton for their generous hospitality and for many memorable days spent at Sirocco on Lake Naivasha, at their elephant archive in Nairobi and at the Samburu Elephant Research Project. I am also grateful to Cynthia Moss, Joyce Poole and staff at the Amboseli Elephant Research Project. My own encounters with elephants have taken place in many different parts of Africa over the years: in the Zambezi Valley, Hwange, Chobe, the Skeleton Coast, the Luangwa Valley, Manyara, Tsavo and the Masai Mara. But nothing has surpassed the enjoyment of watching elephants close up in Samburu and in Amboseli in the company of researchers knowledgeable about their family history and habits. My thanks are also due to Dave Cumming, Karen McComb, Steve Cobb, Juliet Brightmore, Felicity Bryan, Rachel Whitehead, and Peter Osnos and the staff of PublicAffairs.

# I

## *Manyara*

---

At the foot of a waterfall that cascades down the escarpment of the Great Rift Valley into the Ndala River, there is a pool with sandy edges where elephants like to congregate. Overlooking the pool stands a high bank with views stretching across canopies of flat-topped acacia trees towards Lake Manyara a few miles beyond. It was here, in 1966, that a young Scottish biologist, Iain Douglas-Hamilton, decided to set up camp for what was to become a pioneering study of African elephants.

Douglas-Hamilton's initial assignment was to assess the damage to trees caused by elephants in Manyara, a small national park in northern Tanzania lying between the Rift Valley escarpment and the lake. It contained one of the highest densities of elephant population ever recorded: nearly 500 elephants were estimated to be living in an area of no more than thirty-five square miles. They had recently started to strip bark from acacia thorn trees, destroying whole tracts of woodland. No one knew what they needed from the bark nor what the consequences of the damage they inflicted on the acacia forest would be. The park authorities also wanted to learn more about an elephant migration that occurred in the dry season when large numbers disappeared into a huge cloud forest on top of the escarpment.

Soon after his arrival, Douglas-Hamilton realised the

need for a more thorough, long-term approach to under-standing the impact of elephants on their environment in Manyara. Only by ascertaining their birth rates and death rates and their movements in and out of the park would he be able to gain a clear perspective on the scale of the problem. This in turn meant that he would have to acquire the ability to recognise by sight a large number of individual elephants. No one had hitherto undertaken a study of individual elephants in the wild.

Day after day he followed elephants, taking photo-graphs for identification, sometimes on foot, or crouching in trees, or from a battered Land-Rover. The most effective method of identification, he discovered, was to memorise the shape of their ears which were usually marked by notches, slits or holes. Another distinguishing feature was the pattern of veins on their ears which provided as accurate a means of identification as human fingerprints.

The work was hazardous. Each elephant had to be photographed with its ears spread out, facing the camera, a position most commonly adopted when an elephant was alarmed and about to charge. Douglas-Hamilton was charged on countless occasions, making many narrow escapes. Once, following an elephant trail up the escarp-ment, he was trampled by a rhinoceros and seriously injured.

But gradually the elephants grew accustomed to his pres-ence, accepting him as harmless, and Douglas-Hamilton was able to observe elephant life at close hand undisturbed for hour after hour. He became familiar with their different personalities, naming many of them after friends or literary and historical figures, instead of by scientific numbers.

His cast of characters included: Boadicea, a leading

matriarch who made fierce threat charges but never carried them through; the ferocious Torone sisters who charged without warning and in total silence; and Virgo, a gentle elephant, intensely curious, who would walk up to Douglas-Hamilton, standing only a few feet away, eventually allowing him to touch the lip of her extended trunk. After four years in Manyara, Douglas-Hamilton had named or numbered almost all the elephants in the park.

By observing them so closely, he gained new insights into their remarkable social life, finding many resemblances to human behaviour. The basis of elephant society, he concluded, was the family unit, led by a matriarch and consisting of sisters and cousins, with their various offspring. Family members were bound by strong ties of loyalty which lasted for life. Cows were devoted to the well-being and protection of their offspring, exercising parental care well into their early teenage years. In times of distress and danger they held fast to their family ties and swiftly combined to ward off threats by predators. Family units in turn were linked to wider 'kinship' groups, usually related, whose company they often shared. In all, Douglas-Hamilton discovered some forty-eight cow-calf family units in Manyara, most of them belonging to larger kinship groups.

To help him keep track of elephant movements, he devised a method of fixing radio collars carrying a transmitter around the neck of an elephant that had been temporarily immobilised by drugs. The first experiments were successful. For twenty days and nights he was able to stay on the trail of a young bull as he wandered up the escarpment, along densely forested gorges and down to the lakeside, recording his eating and drinking habits and the company he kept. As a next step, Douglas-Hamilton acquired a light aircraft, enabling him to pick up radio signals in the air from a distance as far as thirty miles away, providing far greater scope to plot elephant movements.

It was on a trip to Nairobi in 1969 to get his plane serviced that he met a girl with long dark hair and a vivacious nature named Oria Rocco, to whom he was immediately attracted. The daughter of an Italian aviator and a French sculptress, she had been brought up on the family farm on the shores of Lake Naivasha in Kenya. She was swiftly enticed to Douglas-Hamilton's camp on the Ndala River and soon immersed in the world of elephants. Their first daughter, Saba, was born the following year.

Shortly before leaving Manyara in 1970, the Douglas-Hamiltons introduced their three-month-old daughter to Virgo who was in the company of her close relatives. Standing close to Oria, Virgo moved the tip of her trunk over Saba in a figure of eight, smelling her gently. Oria recalled the moment: 'We both stood still for a long while, facing each other with our babies by our sides.'

Douglas-Hamilton's work in Manyara, however, was overtaken by disaster. During the 1970s and 1980s, as ivory prices on the world market soared, poachers armed with automatic weapons decimated many of Africa's elephant populations. Between 1979 and 1989 the total elephant population was halved.

Manyara's elephants were spared the ravages of the 1970s, but not the next onslaught of the 1980s. In 1987 Douglas-Hamilton returned to Manyara to find out what had happened. From the air, the old elephant haunts seemed deserted. There were dead elephants everywhere. On his last day, he climbed a tree at Ndala to watch a line of approaching elephants. Among them was Virgo, accompanied by a few stragglers and orphans. Fifty yards away, she caught the whiff of human scent, reared her head, spun round and fled down the hill in terror. 'I knew then that Manyara's elephant society had suffered a pogrom,' said Douglas-Hamilton.

———————

Elephants have fascinated mankind since early civilisation. They have been used as symbols of wisdom and power. They have featured in myths and religions, representing prudence, constancy and many other virtues. They have appeared on coins, in architecture, sculpture and painting, in folklore and nursery tales. Ancient writers dwelt on their intelligence, their capacity for learning and their gentle character. They noted too their unusual sense of death. Modern biologists have discovered they possess one of the most advanced and harmonious social organisations in the world of mammals.

Yet, for all their talents and their endearing nature, African elephants have been among the most persecuted animals on earth. The ivory they carry has been prized since ancient times for its unique beauty and sensual qualities. So relentless has been the demand for ivory that many of Africa's great herds have been driven to extinction. After decades of slaughter in the nineteenth century, governments in Africa set aside vast areas of land as national parks and wildlife reserves to ensure the survival of endangered species such as elephants. But even there, they were not safe. In the late twentieth century, just as scientists were beginning to understand the complexities of elephant society, a new onslaught started. Among the victims were many of Manyara's elephants—the first individuals in the wild that scientists had ever studied. The onslaught was so systematic that biologists warned that if it continued African elephants would die out altogether. Yet, year after year, their warnings went unheeded.

# 2
## The Land of Punt

Elephants once roamed over the whole of Africa, from the shores of the Mediterranean in the north to the slopes of Table Mountain in the south. They thrived even in the vast expanses of the Sahara, as ancient rock paintings there testify. But in the third millennium BC, as the climate changed and the great rivers and lakes of the Sahara gradually dried up, elephant and man alike retreated before the encroaching desert, some moving northwards towards the Mediterranean, others withdrawing to the *sahel*, the 'shore' of the Sahara, lying hundreds of miles to the south. The elephants' range in northern Africa was thus cut in two. By 2000 BC, the desert had become almost empty, a landscape of bare rock and moving mountains of sand eventually covering nearly one-third of the continent. In the central Sahara, only the rock paintings remained.

Elephants survived in Egypt for a time. The first pharaohs enjoyed hunting them and coveted them for their ivory. Some were captured and tamed. As early as 3000 BC the Egyptians had developed different hieroglyphs to distinguish between wild elephants and trained ones. But in Egypt too, as the climate became increasingly arid, the local elephant population dwindled and disappeared, along with the rhinoceros and the giraffe.

The pharaohs thus turned their attention to other areas, eastwards to Syria where herds of Asian elephants could

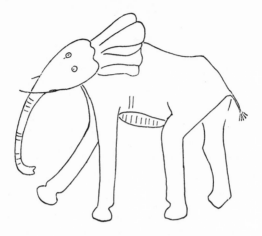

still be found, and southwards, along the Nile Valley to Nubia and beyond, to a region known as the land of Punt, renowned for its riches in ivory, incense, timber, gold and other minerals.

Large-scale expeditions were despatched up the Nile to acquire 'the marvels of the land of Punt'. The first recorded expedition set off during the reign of the pharaoh Sahure in the twenty-fifth century BC. In the twenty-third century BC an Egyptian nobleman named Harkhuf, the governor of Elephantine (Aswan), led four expeditions there.

Other expeditions were sent to Punt by ship, along the Red Sea coast. In the fifteenth century BC Queen Hatsheput ordered the construction of a fleet of five ships on the Nile which were then carried across the desert from Thebes to the Red Sea coast together with all the provisions and equipment required for the voyage southwards. Among the 'marvels' the expedition brought back were 700 elephant tusks.

Ivory was in constant demand, both in Egypt and in other lands around the eastern Mediterranean. Since the fifth millennium BC it had been treasured as a symbol of wealth and status. Its subtle glowing colour and sensual surface appealed to carvers and the rich elite alike. The Egyptians first used it to make luxury items such as combs, bangles and pendants; then they devised more elaborate products such as statuettes, ornaments, chests and gaming boards. Pharaohs' graves were filled with carved ivory objects and inlaid ivory furniture to accompany them on their way to the afterlife. Tutankhamun was provided with an exquisite ivory headrest.

The ivory-carving trade spread throughout the eastern Mediterranean, much of it carried on by itinerant Phoenician craftsmen. Ivory workshops flourished in Crete, in Cyprus and in the Greek city of Mycenae.

The use of ivory became ever more extravagant. The Old Testament records how in *c.*1000 BC King Solomon ordered the construction of a 'great ivory throne' overlaid with gold. 'The like of it was never made in any kingdom,' says the Book of Kings. Solomon's temple in Jerusalem also used

huge quantities of ivory. King Ahab later built a palace that was so heavily ornamented in ivory that it was known as the ivory house. Indeed, in the Hebrew kingdom, ivory became synonymous with luxury and decadence, prompting the Old Testament prophet, Amos, to warn, 'The houses of ivory shall perish.'

The Greeks, in the fifth century BC, developed a similar passion for ivory. They took particular delight in a type of statuary known as chryselephantine in which ivory repre- sented the flesh of a figure while gold was used for robes and hair. Using ivory and gold, the Greek sculptor Phidias built statues of Athena nearly forty feet high and of Zeus more than fifty-eight feet high.

So great was the demand for ivory from ancient civilisa- tions that by 500 BC the Syrian herds had all been driven to extinction.

The land of Punt thus became increasingly important to the ivory trade. Soon after the Ptolemies, a Greek dynasty, took power in Egypt in 323 BC, they organised new expedi- tions to exploit the riches of the south—an area they called Ethiopia, 'the land of the blacks'. In about 270 BC the second Ptolemy, Philadelphus, sent a mission 'to investigate about hunting elephants' and to found a new settlement on the Red Sea coast he called Philotera, about 100 miles south of Suez. A chain of hunting stations was established further south along the coast eventually reaching as far as Bab el Mandeb, opposite Aden. Many stations bore the names of expedition leaders: the Island of Straton; the Look-out Post of Demetrius; the Altars of Conon; the Harbours of Antiphilus.

There were plenty of ivory stocks and elephant herds to be found. The Greek historian Polybius reported in the second century BC that 'in the outlying parts of the province, where it marches with Ethiopia, elephants' tusks serve instead of doorposts in the houses, and partitions in these buildings and in stabling for cattle are made by using elephants' tusks for poles.'

Hunting parties were organised with the help of local Ethiopian elephant fighters known as Elephantomachoi who used hamstringing as a common method of attack. An Alexandrian geographer, Agatharchides of Cnidus, writing in the second century BC, provides a graphic account of how the Elephantomachoi pursued their quarry.

Hiding in a tree, a hunter waited until an elephant passed beneath. Then:

He seizes its tail with his hands and plants his feet against its left thigh. He has hanging from his shoulders an axe that is light enough for a blow to be struck with one hand and extremely sharp. Grasping this in his right hand, he severs the right hamstring tendon by striking repeated blows while supporting his own body with his left hand. They apply themselves with remarkable swiftness to these tasks as though their own soul had been set before each as a prize. For the situation allows for no other outcome than that he should subdue the beast or himself die.

Sometimes, as the hamstrung creature is unable to turn because of its impaired movement, it settles back on the wounded spot, falls and kills itself and the Ethiopian, or sometimes it squeezes the man against a rock or a tree and crushes him with its weight until it kills him.

Some elephants, maddened with pain, make no attempt to fend off their assailant but take flight across the plain until their attacker by repeatedly striking at the same spot with his axe severs the tendon and renders the beast helpless. When the animal falls, the men run to it as a group; and while it is still alive, they cut pieces of meat from its hindquarters and feast.

Another method that local hunters used, according to Agatharchides, was the bow and arrow.

Three men with one bow and many arrows coated with snake venom lie in wait in the bush beside the animals' trails. When a beast approaches, one man holds the bow, bracing it with his foot, and two other men employ all their strength to draw the bowstring and release the arrow. Their sole target is the middle of the flank with the result that the arrow pierces through its skin and cuts and wounds its stomach. The huge beast, struck and mortally wounded, loses its strength and collapses.

While the ivory supplies were highly valued, however, the Ptolemies were interested in Ethiopia's elephants not only for their ivory. Facing challenges from rival dynasties in the Levant, they wanted them for war purposes. Orders went out to capture live elephants.

# 3
## Arms and the Elephant

When Alexander the Great led his Macedonian army into its final battle to conquer the Persian Empire in 331 BC, he was disconcerted to find among King Darius's army a contingent of fifteen elephants decked out in armour and ready for combat. Alexander had never previously encountered elephants.

On this occasion, they gave him little trouble. But four years later, after invading northern India, Alexander met a far more formidable force. Drawn up against him on the east bank of the River Hydaspes was a huge Indian army protected in front by a battle line of 200 heavily armoured elephants. Some carried towers on their backs manned by soldiers armed with long pikes. The mere sight and smell of the elephants was enough to terrify Alexander's cavalry horses and alarm his men. In battle, they wrought terrible havoc, impaling men on their tusks, seizing them with their trunks and dashing them to the ground. But eventually Alexander won a famous victory.

He was sufficiently impressed by the fighting abilities of Indian elephants to incorporate them into his own army. Against soldiers and horses that had never seen them, they created panic and pandemonium. Elephants became the fore-runner of the modern tank, trained to intimidate infantry, to launch cavalry charges and to tear down fortifications. They were also used to execute prisoners.

After the death of Alexander in 323 BC, his empire and his elephant corps were divided among his generals. In the eastern Mediterranean two rival dynasties emerged: the Ptolemies, who controlled Egypt and Palestine from their metropolis at Alexandria, at the mouth of the Nile; and the Seleucids who held northern Syria, Mesopotamia and Persia. In a long series of wars fought over possession of southern Syria during the third century BC, both sides used war elephants. But while the Seleucids were able to obtain replacements from India, the Ptolemies' access to India was blocked. Their elephant corps was soon depleted. They therefore turned to Africa for an alternative source of supply.

A new base to capture elephants was established on the Red Sea coast at a point near the Baraka River, about fifty miles south of Port Sudan. Named Ptolemais Theron, or Ptolemais of the Hunts, it grew into 'a great city', according to a contemporary inscription, self-supporting in crops and cattle.

But local Ethiopian hunters proved unwilling to help capture elephants alive, as Agatharchides noted: 'Ptolemy urged the hunters to refrain from killing elephants in order that he might have them alive ... Not only did he not persuade them but they said that they would not change their way of life in exchange for the whole kingdom of Egypt.'

Nevertheless, the enterprise at Ptolemais Theron eventually succeeded. 'Elephants were caught in great number for the king and brought as marvels to the king, on his transports on the sea.'

The voyage to Egypt in specially constructed transport ships was hazardous. Crews had to deal with treacherous head winds, hidden coral reefs and the constant risk of shipwreck. The Greek writer Diodorus records:

The ships which carry the elephants, being of deep draught because of their weight and heavy by reason of their equipment, involve their crews in great and terrible dangers.

Since they run under full sail and often are driven before the force of the winds during the night, sometimes they strike the rocks and are wrecked, at other times they run aground on slightly submerged spits.

The sailors cannot go over the sides of the ships because the water is deeper than a man's height, and when in their efforts to rescue their vessel by means of their punt-poles they accomplish nothing, they jettison everything except their provisions.

At first, elephants were taken by ship all the way to the head of the Gulf of Suez, 1,000 miles away, and from there by canal to Memphis. But so dangerous was the long sea route that a new port was established for them halfway along the coast at Berenice Trogodytica. From Berenice, the elephants walked overland through the Eastern Desert to the Nile along a caravan route specially equipped with camps and watering places. Their eventual destination was the main elephant stables at Memphis. Some were taken to Alexandria for display in a zoo which Philadelphus established there.

African elephants were first deployed in battle against the Seleucids during the Third Syrian War in the 240s BC, in a campaign that the Ptolemies easily won. During the Fourth Syrian War, however, they fared less well.

At the Battle of Raphia in Palestine in 217 BC, African and Asian war elephants met in a decisive encounter. Both sides used towers containing soldiers. But Ptolemy's African elephants were smaller in size than their opponents; they came from the 'cyclotis' subspecies of elephant, or forest elephant, which then inhabited northeast Africa, standing no more than eight feet tall at the shoulder. They were also heavily outnumbered. Ptolemy fielded seventy-three elephants; his enemy, Antiochus, one hundred and two.

Antiochus opened the battle, ordering a charge of sixty elephants on his right wing against the forty on Ptolemy's left wing. The Greek historian Polybius depicted the ensuing clash: 'Only some few of Ptolemy's elephants came to close quarters with their opponents, and the men in the towers on the back of these beasts made a gallant fight of it, lunging with their pikes at close quarters and striking each other, while the elephants themselves fought still more brilliantly, using all their strength in the encounter and pushing against each other, forehead to forehead.'

He described their fighting style: 'With their tusks firmly interlocked and entangled they push against each other with all their might trying to force the other to give ground, until the one who proves the strongest pushes aside the other's trunk, and then, when he has once more made him turn, he gores him with his tusks.'

Some of Ptolemy's elephants, he said, engaged bravely in duels. But most, he claimed, were afraid to join battle. 'Unable to stand the smell and the trumpeting of Indian elephants, and terrified, I suppose, also by their great size and strength, they immediately run away.'

Ptolemy's left wing was driven back in confusion. On his right wing, meanwhile, his elephant contingent, outnumbered by thirty-three to forty-two, refused to engage their opponents. While the outcome of the battle was victory for Ptolemy, his African elephant corps noticeably failed to distinguish itself; and the Ptolemies lost all enthusiasm for using them in warfare again.

---

Elsewhere in Africa, however, African war elephants were still highly prized. During the long struggle for supremacy in the western Mediterranean between Carthage and Rome, the Carthaginians used them time and again, convinced of their military value. The Carthaginian general Hannibal was a notable advocate of elephant warfare; and his African war elephants achieved lasting fame. But just as much as they contributed to his achievements, so too they figured prominently in his eventual defeat and downfall.

From their base in Carthage, on the Gulf of Tunis, the Carthaginians had ready access to large herds of elephants which populated the coastal plains from Tunisia to Morocco and the forests at the foot of the Atlas Mountains beyond. One of the first written records of African elephants came from a Carthaginian explorer named Hanno who, in the early fifth century BC, set out on a pioneering journey beyond the Pillars of Hercules—the Straits of Gibraltar—to establish Punic settlements on the west coast of Africa. After passing Cape Soloeis (modern Cape Cantin), he came across marshes near the foot of the Atlas Mountains which, he said, 'were haunted by elephants and multitudes of other grazing beasts'.

In the third century BC the Carthaginians were quick to learn the techniques of adapting elephants as war machines. Within the walls of Carthage itself, stables were built, it was said, for as many as 300 elephants. The elephant corps was soon deployed in earnest. In 262 BC the Carthaginians sent a detachment of fifty elephants across the Mediterranean to Sicily to participate in the First Punic War against the Romans. Their tactics of combining cavalry and elephants proved highly effective.

But it was the young general, Hannibal, born in 247 BC, who engaged in the most daring exploit of all. From his base in Spain, Hannibal devised plans to take the Romans by surprise by marching an army 1,500 miles overland, across the Pyrenees, into the unknown lands of France, over the high passes of the Alps, and through northern Italy to threaten Rome itself. A key part of his army was an elephant corps numbering thirty-seven. Hannibal expected that Roman forces, unprepared for an elephant attack, would retreat in disarray.

Setting out in the summer of 218 BC, later than planned, Hannibal's army encountered unexpected resistance from local tribes in Catalonia, thus delaying its progress. It took some four months for the elephant corps, consisting almost entirely of smaller 'forest' elephants from North Africa, to climb the Pyrenees, plough through the marshes of southern France, and reach the River Rhône.

Crossing the Rhône was particularly hazardous. The point Hannibal chose was between Fourques and Arles where the river runs in a single stream about 300 yards wide. For an army of 38,000 infantry, 8,000 cavalry and thirty-seven elephants, the crossing on its own was a major feat. But Hannibal's troops also faced attacks from Gauls on the far bank; and they knew that a Roman army was fast approaching.

Elaborate arrangements had to be made to get the elephants across. Engineers built huge piers out into the river; then fastened rafts alongside and covered them with an

even layer of earth to encourage the elephants to embark. Led by their drivers, two cow elephants were taken on to the rafts; others followed obediently.

But when the rafts were cast off to be towed to the far bank, pandemonium broke out. According to Polybius, the elephants rushed wildly about trying to get back to land. Eventually, most calmed down. A few, however, flung themselves into the river. And although they managed to wade ashore safely, their drivers drowned.

With the elephant corps in the rear, Hannibal's army marched up the Rhône Valley for ten days, then turned eastwards to begin the ascent towards the Alps, finding their way through unmapped mountain gorges and across swollen rivers. Hostile Gaul tribes harassed the long columns as they climbed high into the mountains, but never dared approach the elephants, fearful of their strange appearance.

After nine days Hannibal reached the highest pass,

beneath Monte Viso, nearly 10,000 feet above sea level. Below he could see the plains of the River Po. But the descent there was to be even more perilous.

As his army paused to rest on the barren heights, tired, cold and short of food and fodder, it began to snow. The way down became slippery and treacherous. Men, horses and pack-animals slid over precipices and perished in their hundreds from exposure and exhaustion. Corpses littered the way. Not far from the summit, a landslide made the track too narrow for elephants and pack-animals to pass. For three days, while the elephants waited wretched from hunger, Hannibal's soldiers laboured to widen the track, lighting fires to crack the rockfall.

After fifteen days crossing the Alps, Hannibal succeeded in reaching the plains below. He had lost nearly half of his army—some 20,000 foot soldiers and horsemen. But all thirty-seven elephants survived.

The sudden appearance of Hannibal's army in northern Italy struck the Romans like a thunderbolt. They hastily organised forces to oppose him. But in one battle after another they were defeated. At Trebia, in the winter of 218 BC, fighting amid the sleet and snow, Hannibal's elephants caused havoc among the Roman cavalry. News of the Roman defeat at Trebia brought panic to the city of Rome.

But elephant casualties steadily mounted. Crossing the Appennines in atrocious weather, seven starved to death. Others perished in the flooded marshes along the Arno Valley. Hannibal himself, suffering from an eye infection, rode the last surviving elephant to keep himself above the swirling waters, but still lost the sight of one eye.

Hannibal's army roamed about Italy for fifteen years. He

reached the gates of Rome but failed to take the city. Elephant reinforcements were sent from Spain and from Carthage. But the Romans by then had become skilled in the techniques of dealing with enemy elephants. In military training, soldiers were taught to attack their trunks to inflict great pain. In combat they wore spiked armour, drove javelins at them, hamstrung them with axes, used fire and flaming arrows and trumpet blasts. In terror and confusion, elephants would often turn back, causing mayhem within the ranks behind them.

Nine years after Hannibal crossed the Alps, his brother Hasdrubal set out from Spain to reinforce him with a new army and a new African elephant corps, taking the same perilous route through the mountains. But this time Roman armies were lying in wait. And at the Battle of Metaurus in 207 BC, Hasdrubal's elephants let him down.

'As the din increased and the fighting grew hotter,' wrote the Roman historian Livy, 'the [elephants] got out of control, charging this way and that like rudderless ships between the two lines as if they did not know to whom they belonged.'

To prevent further chaos, their own riders killed them.

'More of the elephants were killed by their own riders than by the enemy,' wrote Livy. 'The riders used to carry a mallet and a carpenter's chisel and when one of the creatures began to run amuck and attack its own people, the keeper would put the chisel between its ears at the junction between head and neck and drive it in with a heavy blow. It was the quickest way that had been found to kill an animal of such size once it was out of control.'

The Battle of Metaurus sealed the fate of the Carthaginian attempt to defeat the Romans in Italy. On that day, the balance of power in the Mediterranean shifted irrevocably.

In 204 BC, when the Romans sent an expeditionary force to North Africa to seize Carthage, Hannibal was obliged to withdraw from Italy to defend his African homeland.

The decisive battle was fought at Zama in 202 BC. Once again, Hannibal depended heavily on his elephant corps, opening the battle by sending eighty elephants charging into Roman lines. But, terrified by the blare of bugles, some turned against their own side, others rampaged through gaps the Romans made in their lines and were speared to death.

Hannibal conceded defeat. One of the terms of peace dictated by the Romans was that the Carthaginians were required to surrender all their elephants and undertake not to train any more for military purposes.

Now, a new threat faced Africa's elephants, this time from Rome.

# 4
## Roman Games

---

All of northern Africa, including the elephant ranges, eventually fell under Roman control. In 146 BC the Romans annexed Carthage's territory, giving it the name of Africa. In 46 BC they annexed the kingdom of Numidia, further west along the coast, calling it Africa Nova. One hundred years later, they annexed the territory west of Numidia, then known as Mauretania, which stretched as far as the Atlantic coast.

The Romans valued elephants not so much for military purposes as for public spectacle and entertainment. African elephants first appeared in Rome during the third century BC in triumphal processions organised by military leaders to mark their victories. During the second century BC they became a popular feature among the displays of exotic animals paraded at Roman games. They were later trained as combatants to fight each other and other animals, such as bulls and rhinoceroses. Then they were pitted against gladiators armed with lances and glowing firebrands.

The spectacles became increasingly extravagant and bloody. In the opening days of the games organised by the Roman general Pompey in 55 BC, no fewer than 600 lions were massacred. On the final day it was the turn of the elephants. A group of some twenty elephants was led into the Circus to fight Gaetulians from Africa armed with javelins. The Roman historian Pliny described how, much to the delight of the crowd, one elephant put up a heroic fight. Wounded in its feet, it crawled on its knees towards the Gaetulians, snatching their shields and tossing them into the air. Another elephant was killed by a single blow from a javelin which struck it just below the eye. The remaining elephants then tried to escape by breaking through the iron barriers of the enclosure protecting spectators. When their attempt failed, they stood in the arena waving their trunks in desperation and trumpeting piteously. So distressed did the crowd become at their plight, according to Pliny, that they rose to their feet and cursed Pompey for his cruelty.

'The result,' said the writer Cicero, who witnessed the occasion, 'was a certain compassion and feeling about the affinity between men and elephants.' And he asked: 'What pleasure can a man of culture find in seeing a noble beast run through by a hunting spear?'

The games, however, remained popular. At Caesar's games in 46 BC, one contest consisted of a group of twenty elephants set against 500 infantry; it was followed by a second contest in which twenty elephants, with towers on their backs holding three armed men, engaged 500 infantry and cavalry. Under the emperors Claudius and Nero, combat between elephants and men single-handed became the crowning achievement of a gladiator's career. Elephants were also used as executioners when men and women prisoners were condemned *ad bestios* and dragged at dawn into the arena to be killed.

Despite the brutality of the games, Roman leaders prized elephants and continued to mark their military victories with elephant displays. Returning from Africa in triumph in 80 BC, Pompey decided to enter Rome in a chariot drawn by four elephants and he was only prevented from doing so

because the city gate proved too narrow. In 46 BC Caesar celebrated his African victory by organising a procession to his home accompanied by forty elephants carrying torches in their trunks. A medal produced in 18 BC to mark the death of the emperor Augustus depicts a statue of him standing god-like in a chariot being drawn by four African elephants waving their trunks in acclamation.

Elephants were much admired for their ability to learn tricks and for their dexterity in performing them. In the arena, elephants were brought to kneel before the imperial box and trace Latin phrases in the sand with their trunks. On one occasion they were said to have bowed and prayed before the emperor before making the sign of the cross with their trunks.

They were taught to participate in social functions and in musical events, to dance gracefully and tap out rhythms with their feet. The historian Aelian described how, at an entertainment organised by the Roman general Germanicus in about AD 12, a group of twelve elephants, wearing the flowered dresses of dancing girls, entered the theatre, formed into line and wheeled about in circles, keeping time in a rhythmic dance and sprinkling flowers delicately on the floor. They were then taken to an elaborate banquet spread out on tables of ivory and cedar where, reclining on couches, they proceeded to eat and drink with the utmost decorum, much to the delight of spectators.

They also learned to walk on tightropes. At a festival that the emperor Nero held in honour of his mother in AD 59, an elephant was led to the highest gallery of a theatre; then it skilfully walked down on ropes to the ground floor, carrying a rider. Pliny tells the story of how one conscientious elephant, frequently scolded for failing to master its tricks, was found at night putting in some extra practice, all by itself, by moonlight.

Rome's liking for elephants meant that the North African herds faced constant raids. But even more perilous was the insatiable Roman demand for ivory. Ivory was used to decorate temples and palaces; carried in triumphal processions; and made into a vast range of luxury goods: thrones, chests, statues, chairs, beds, book-covers, tablets, boxes, birdcages, combs and brooches. Caesar rode in an ivory chariot; Seneca possessed 500 tripod tables with ivory legs; Caligula gave his horse an ivory stable. Consuls and magistrates adopted ivory for their insignia of office, their sceptres and their curule chairs and they sent inscriptions of their appointments to dignitaries and friends on ivory diptychs.

Ivory carvers strove to ever greater heights of achievement. The Roman poet Ovid tells the tale of one sculptor who fell in love with his own creation—the perfect rendering of a woman carved in ivory. The idea was reworked in the twentieth century by the Irish dramatist George Bernard Shaw in his play *Pygmalion*, which later formed the basis of the popular musical, *My Fair Lady*.

Some of Rome's ivory came from India; some from Ethiopia and the Red Sea coast. The kingdom of Aksum in the Ethiopian highlands became a major trade partner, supplying ivory through its port at Adulis. Ethiopia still boasted huge herds of elephant. A Roman envoy travelling to Aksum in the mid-sixth century reported seeing a herd some 5,000 strong on the road from Adulis.

But North Africa suffered the brunt of Roman demand. Early signs of the elephants' plight there were soon evident. In AD 77 Pliny complained about the shortage of African ivory: 'An ample supply of ivory is now rarely obtained except from India, the demands of luxury having exhausted all those in our part of the world [North Africa].'

The Roman Empire in North Africa survived until the fifth century. But by then the North African herds were on their way to extinction.

# 5
## Protégés of the Gods

Throughout the classical world, whatever purpose they served, elephants were accorded high status. They featured in religious ceremonies and victory parades. They appeared on coins, in sculptures and paintings, symbolising wisdom and power. They were said to possess pious habits. They were portrayed as protégés of the gods, the draught-animals of their chariots.

According to Pliny, the Roman historian:

The elephant is the nearest to man in intelligence: it understands the language of its country and obeys orders, remembers duties that it has been taught, is pleased by affection and by marks of honour; nay more, it possesses virtues rare even in man, honesty, wisdom, justice, also respect for the stars and reverence for sun and moon.

Authorities state that in the forests of Mauretania [northwest Africa], when the new moon is shining, herds of elephants go down to a river named Amilo, and there perform a ritual of purification, sprinkling themselves with water, and after paying their respects to the moon return to the woods carrying before them those of their calves that are tired.

Pliny the Elder was so intrigued by elephants that he devoted thirteen consecutive chapters to them in his main literary work, *Historia Naturalis*, which appeared in AD 77. He wrote about their gentle character: 'They never do any harm unless provoked, even though they go about in herds, being of all animals the least solitary in habit.' He was impressed by their cooperative nature: 'Africa captures elephants by means of pitfalls; when an elephant straying from the herd falls into one of these, all the rest at once collect branches of trees and roll down rocks and construct ramps in an attempt to get it out.'

When compiling information about elephants, early writers such as Pliny relied heavily on the work of Aristotle, the father of elephant science, who provided the first lengthy treatise of their biology and natural history in Greece in the fourth century BC. Aristotle's work included discussions on anatomy, reproduction, diet and behaviour. He referred to the elephant's high intelligence: 'the beast which passeth all others in wit and mind'. He noted its peaceable nature. And he spoke eloquently about the trunk: 'The elephant uses its nostril as a hand; for by means of this organ it draws objects towards it, and takes hold of them, and introduces its food into its mouth, whether liquid or dry food, and it is the only living creature that does so.'

SOLI DEO

Much of Aristotle's work was remarkably accurate. But some observations he made led to enduring myths. On their breeding habits he recorded: 'Elephants copulate in lonely places, and especially by riversides in their usual haunts.' Four hundred years later, Pliny repeated the idea: 'Owing to their modesty, elephants never mate except in secret.'

Like his predecessors, the Roman writer Aelian, in the third century AD, devoted much space to stressing the virtuous qualities of elephants. They never abandoned the weak or the young, even when being hunted, he wrote; and they helped the old out of pits into which they had fallen. He recounted 'Ethiopian' tales which told how elephants would not pass by a dead elephant without casting a branch or some dust on the body.

He remarked on their 'uncanny and astounding dexterity' and their ability to learn and perform tricks:

For the present, I intend to speak of their sense of music and their readiness to obey, and their aptitude for learning things which are difficult even for mankind, to say nothing of so huge an animal and one hitherto so fierce to encounter. The movements of a chorus, the steps of a dance, how to march in time, how to enjoy the sound of flutes, how to distinguish different notes, when to slacken pace as permitted or when to quicken at command—all these things the elephant has learned and knows how to do and does accurately without making mistakes. Thus, while nature has created him to be the largest of animals, learning has rendered him the most gentle and docile.

As well as recording the feats that elephants performed for man when tamed, classical writers dwelt on their everpresent fear of man in the wild.

It is said [wrote Pliny] that when an elephant accidentally meets a human being who is merely wandering across his track in a solitary place, it is good-tempered and peaceful and will actually show the way; but that when on the other hand it notices a man's footprint before it sees the man himself it begins to tremble in fear of an ambush, stops to sniff the scent, gazes round, trumpets angrily, and avoids treading on the footprint but digs it up and passes it to the next elephant, and that one to the following, and on to the last of all with a similar message, and then the column wheels round and retires and a battle-line is formed.

Aelian gives a graphic description of elephants charging hunters: 'The animals attack, their ears in passion spread wide like sails, after the manner of ostriches which open their wings to flee or attack. And the elephants bending their trunk inward and folding it beneath their tusks, like the ram of a ship driving along with a great surge, fall upon the men in a tremendous charge, overturning many and bellowing with a piercing shrill note like a trumpet.'

Like earlier writers, Aelian believed that elephants were beloved by the gods. In Mauretania, he wrote, old elephants were said to withdraw to a sanctuary deep in a forest at the foot of the Atlas Mountains protected by the gods, a place that no hunters dared approach. When a king sent 300 picked men to acquire the heavy tusks of old elephants, he said, all but one was struck down by pestilence. Aelian also recorded how, when dying of wounds, elephants would throw up dust to heaven, protesting to the gods about the injustice of their fate.

But man was not their only enemy. A common story in the classical world concerned the struggle between elephants and serpents. In Ethiopia, said Aelian, some snakes were known as 'elephant-killers' and reached 180 feet. Pliny claimed that snakes drank the blood of elephants, draining them dry. The poet Lucan, who had a fascination for African snakes, believed that they could crush elephants. A mosaic in Carthage depicts an elephant in the grip of a python.

What lay behind such tales was the age-old obsession between good and evil. It was the elephant that symbolised light and life and victory over darkness and death. This theme was picked up by medieval Europe where elephants had never been seen.

# 6

## *The Elephants' Graveyard*

On the maps of Africa drawn up by early geographers, the lands beyond the Nile and Ethiopia were marked simply: *Terra incognita*. When the Greek historian Herodotus visited Egypt in about 430 BC, travelling up the Nile as far as the first cataract at Elephantine (Aswan), he could find no one who knew anything about the river's source. A scribe at the sacred treasury of Athene in the city of Sais suggested that the Nile arose from 'fountains' somewhere in the African interior, but could tell him nothing more.

The only glimpse of this vast hinterland for centuries came from a Greek merchant named Diogenes. He claimed that as he was returning home from a visit to India in the middle of the first century AD he had landed on the African mainland at a place called Rhapta—somewhere on the present Tanzanian coast—and then travelled for twenty-five days inland. He arrived, he said, 'in the vicinity of two great lakes, and the snowy range of mountains whence the Nile draws its twin sources'. A century later, the Alexandrian geographer Claudius Ptolemy incorporated this information into his map of the world, and named the source of the Nile as Lunae Montes, the Mountains of the Moon. For 1,700 years Ptolemy's map remained the only guide to the mystery of the Nile's sources.

Not much more was known at the time about the eastern coastline of Africa. In about 60 AD a Greek merchant in

Alexandria published a mariner's guide to the northern regions of the Indian Ocean. It was entitled *Periplus Maris Erythraei*—a circumnavigation of the Eritrean or Red Sea— but its scope encompassed not only the modern Red Sea, but also the Gulf of Aden, the Arabian Ocean and the Indian Ocean, from the western shores of India to the east coast of Africa.

The main focus of the *Periplus* was the trade routes to Asia. Little attention was paid to Africa: a number of landmarks were identified; a few harbours were listed. Only one port of trade was named: Rhapta. It was described as 'the very last market-town of the continent of Azania', the Greek name then given to East Africa. 'There is ivory in great quantity,' the *Periplus* recorded.

Arab and Persian traders eventually established a string of settlements along the east coast of Africa, calling it the land of Zanj. They chose offshore islands on which to found most settlements, making no attempt to penetrate inland. The island of Zanzibar, taking its name from the word Zanj, became a regular destination. Arab dhows set sail for the African coast during the northeast monsoon from November to March and returned home during the southwest monsoon from April to September.

Their main interest was in African slaves. Slaves from Zanj and from Abyssinia—the Arab name for the Ethiopian region—earned a good price in the Arab world and in the Persian Gulf. They were used to build cities, tend plantations, dig canals and labour in mines. Thousands of African slaves were transported there each year.

But ivory was another valued commodity. Demand for African ivory came not only from the Arab world but increasingly from India and from China. Ivory was commonly used in India to produce bridal ornaments, which, on the death of either marriage partner, had to be destroyed. From the sixth century, India no longer produced enough local ivory to meet its own requirements. African supplies were needed to fill the gap. Indian carvers found, moreover, that African ivory was softer and easier to work than Indian ivory and preferred to use it.

After a few generations, the coastal settlements in Zanj grew more prosperous and secure. Muslim merchants developed thriving harbour towns there, with mosques, palaces and grand residences built from coral stone. A distinctive civilisation took root, acquiring the name of Swahili from the Arabic word for shore or coast.

Arab sea captains probed further south, eventually reaching Sofala, a landing point on the Mozambique coast, only a few hundred miles from goldfields on the plateau of Zimbabwe. A gold trade with Zimbabwe soon flourished, bringing increasing wealth to the towns of Zanj which controlled it.

A Baghdad geographer, Abu'l Hasan 'Ali al-Mas'udi, visited Zanj twice in the tenth century and provided the earliest known account of it:

There are many wild elephants but no tame ones. The Zanj do not use them for war or anything else, but only hunt and kill them. When they want to catch them, they throw down the leaves, bark and branches of a certain tree which grows in their country; then they wait in ambush until the elephants come to drink. The water burns them and makes them drunk. They fall down and cannot get up; their limbs will not articulate. The

Zanj rush upon them armed with very long spears, and kill them for their ivory. It is from this country that come tusks weighing fifty pounds or more.

Most ivory, he said, was shipped to Oman and from there to India and China. 'The Zanj, although constantly employed in hunting elephants and collecting ivory, make no use of ivory for their own domestic purposes.'

Al-Mas'udi also described the perils of travelling across the Indian Ocean to Zanj. 'I have sailed on many seas,' he wrote, 'but I do not know of one more dangerous than that of Zanj.' He listed the captains with whom he had sailed, all of whom had been drowned, he said.

It was from sailors' tales of trading voyages along the East African coast that one of the most enduring legends about African elephants emerged. In *Arabian Nights,* a ninth-century collection of Persian stories, Sindbad the Sailor recounts his adventures while trading 'from port to port, and from island to island' in what was the Sea of Zanj. On his seventh and last voyage, he said, he came across an elephants' graveyard.

It happened, according to Sindbad, after he had been captured by pirates and sold to a rich merchant. The merchant gave him a bow and arrows and ordered him to shoot elephants for their tusks from hiding places in the trees. For two months he managed to kill an elephant every day. Then one morning he found himself surrounded by a herd of angry elephants. They tore down his tree and carried him off on a long march, leaving him on a hillside covered with elephant bones and tusks. He realised, he said, it was an elephants' graveyard and that he had been brought there to be shown there was no need to kill elephants when their tusks could be obtained merely for the trouble of picking them up.

When *Arabian Nights* was translated in Europe in the eighteenth century, Sindbad the Sailor and his adventures became a permanent part of Western folklore.

# 7
## *Nature's Masterpiece*

---

An African elephant arrived on the coast of England in 1254. It was a present for Henry III, King of England, from Louis IX, King of France, which he is said to have acquired during a crusade to Palestine. King Henry ordered elaborate arrangements to be made for its accommodation at the royal menagerie in the Tower of London. 'We command you,' he wrote to the Sheriff of London, 'that ye cause without delay, to be built at our Tower of London, one house of forty feet long and twenty feet deep, for our elephant.'

A Benedictine monk, Matthew Paris, was commissioned to make an illustration of it. His illustration shows the elephant standing fastened to a stake by the ankle and being fed by its keeper.

King Henry's elephant immediately became one of the sights of London. But after only two years in the Tower, it died, possibly as a result of being given too much red wine to drink.

Such sightings of elephants in medieval Europe, however, were rare. Elephants remained largely a mystery, the subject of much myth and speculation. Few people knew what they actually looked like. Elephants depicted in medieval manuscripts and carvings often bore little resemblance to the real animal. Information about the elephant was scarce; much of it was simply borrowed from writers from the classical world, a thousand years before, repeating legends they had heard.

As in classical times, the elephant was held in high esteem. It was said to possess great moral virtues and was used to symbolise the cause of good against evil. Elephant carvings adorned churches and cathedrals across Europe, often depicting a struggle between an elephant and a serpent. Bestiaries written in the Middle Ages conveyed the same message about the elephant's noble character. Writers in the sixteenth and seventeenth centuries carried on the tradition. In the sixteenth century a Swiss naturalist, Conrad Gesner, writing in Latin, enumerated the elephant's many moral attributes. In the seventeenth century a French writer, S. de Priezac, produced a treatise entitled *Histoire Des Éléphants*, praising the elephant as 'a subject in which moral virtues stand out, polity prevails, integrity is triumphant and torment and punishment the sole reward for vice'. Each of his chapters dealt with a particular virtue: moderation, piety, prudence, pride, tact, affection for the young, respect for elders, propriety, charity, mercy, presence of mind, intelligence, fidelity and justice.

An English parson and naturalist, Edward Topsell, included twenty pages on the elephant in his *Historie of the Foure-Footed Beasts*, published in London in 1607. He began by observing: 'There is no creature among al the Beasts of the world which hath so great and ample demonstration of the power and wisdome of almighty God as the elephant.'

Drawing much of his material from classical writers such as Aristotle, Pliny and Aelian, he dwelt on their pious nature:

They have a kind of Religion, for they worshippe, reverence, and observe the course of the Sunne, Moone, and Starres; for when the Moone shineth, they go to the Waters wherein she is apparent, and when the Sunne ariseth, they salute and reverence her face: and it is observed in Aethiopia, that when the Moone is chaunged untill her prime and appearance, these Beastes by a secret motion of nature, take boughes from of the trees they feede upon, and first of all lift them up to heaven, and then looke uppon the Moone, which they doe many times together; as it were in supplication to her. In like manner they reverence the Sunne rysing, holding up their trunke or hand to heaven in congratulation of her rising.

Topsell was impressed by their other attributes: 'There is not any creature so capable of understanding as an Elephant. They are apt to learne, remember, meditate and conceive such things as a man can hardly perform.' He repeated observations made by Aristotle and Pliny about their breeding habits: 'They are modest and shamefast about procreation, for at that time they seek woods and secret places.'

More accurately, he noted the way that death affected

them: 'I cannot omit their care, to bury and cover the dead carcasses of their companions, or any others of their kind; for finding them dead they pass not by them till they have lamented their common misery, by casting dust and earth on them, and also green boughs, in token of sacrifice, holding it execrable to do otherwise.' Topsell also stressed their gentle nature, how they would never fight or strike man or beast unless provoked.

Other writers of the time expressed similar views. In his poem 'The Progress of the Soul', published in 1601, John Donne wrote:

> Nature's great master-peece, an Elephant,
> The onely harmlesse great thing . . .

Medieval Europe also witnessed a revival of the ivory cult. During the thirteenth and fourteenth centuries European sculptors produced a series of masterpieces of Gothic ivory carving, mainly of religious subjects. Among the items on display at the Ste-Chapelle in Paris, which was opened by Louis IX in 1248 to house relics of the Passion he had acquired from Constantinople, was a newly commissioned figure of the Virgin and Child smiling joyfully. It was greeted with immediate acclaim and became one of the most influential ivory carvings ever produced. Workshops flourished in Dieppe, Paris and Soissons, gaining fame for their elegant statuettes, crucifixes, reliquaries and diptychs. Among the masterpieces they produced were a number of miniature private altarpieces, such as the Soissons diptych, now in the Victoria and Albert Museum in London.

There was also increasing demand for ivory items for domestic use, such as combs, mirror cases and caskets. In Paris in the thirteenth century the volume of work for ivory carvers was so great that they formed specialised groups, some concentrating on producing crucifixes and knife handles; others on lanterns, or writing tablets, or dice.

Caskets were a popular item. The wardrobe accounts of Edward I, King of England, for 1299–1300, included an ivory coffer full of personal objects 'for the king's private delight', such as rings, seals, precious stones and purses. Many caskets were ornamented with illustrations from medieval romances.

But ivory was often in short supply. Europe's trade links with Africa and Asia were impeded by the advance of Muslim rule in northern Africa and the Middle East.

A new era of European exploration was about to begin, however, one that would change the course of African history and, with it, the fate of its elephant populations.

# 8

## *Footholds on the Coast*

———

In 1415 a Portuguese armada, carrying the largest army ever assembled by a Portuguese king, crossed the Mediterranean on a new crusade against Islam, aiming to capture the fortress town of Ceuta on the Moroccan coast. On board was Prince Henry, the ambitious twenty-one-year-old son of King John, determined to make his mark as a crusader and hoping that the capture of Ceuta would prove to be only the start of Portuguese military expansion in North Africa.

Ceuta was a valuable prize. It was one of the strongest fortresses in the Mediterranean, guarding its western approaches; it was a major commercial port; and it was a northern terminal of the trans-Sahara caravan trade routes bringing gold and ivory across the desert from the Muslim kingdoms of the Sahel. When Ceuta fell to the Portuguese in a single day in August 1415, its capture was greeted as a great triumph. Portuguese envoys in Europe proclaimed the town to be the 'gateway and key to all Africa'. From wealthy traders captured at Ceuta, the Portuguese learned about the sources of gold shipments crossing the Sahara. Some traders spoke of a 'River of Gold' far to the south flowing into the Atlantic.

But Ceuta remained no more than an isolated enclave on the North African coast, surrounded by Muslim adversaries and dependent on Portugal for supplies. The gold trade lay beyond reach. Henry's attentions turned to other ventures in the Atlantic, to the Madeira Islands, the Canary Islands and the Azores. As Portugal's sea power increased, however, he resolved to find a sea route to the goldfields of Africa.

Hitherto, sailors had ventured no further south along the Atlantic coast of Africa than Cape Bojador, a barren coastal landmark 100 miles south of the Canary Islands, notorious for its fogs and heavy surf. The prevailing wind and current there, running from the north, made return journeys hazardous. Several ships probing southwards had never returned. Beyond the cape lay what medieval geographers knew as the 'Torrid Zone'—a treacherous sea and an inhospitable coastline stretching for hundreds of miles into the unknown. In Arabic, Cape Bojador was known as Bon Khatar: 'Father of Danger'.

Under Henry's direction, Portugal pioneered major advances in shipbuilding and navigation. The Portuguese fleet was equipped with newly designed caravels, which were highly manoeuvrable and ideally suited for reconnaissance along unknown coasts.

Year after year, Henry sent expeditions southwards along the African coast. His aim by now was not only to outflank the trans-Sahara routes and gain direct access to the goldfields but to search beyond for the land of Prester John, a legendary Christian king said to rule over an empire in the interior, cut off from the rest of Christendom by the Muslim powers that controlled North Africa.

The exploration of the West African coast was swift and dramatic. In 1434 a Portuguese crew sailed around Cape Bojador and returned safely against the wind. In 1436 Portuguese mariners reached an inlet 250 miles beyond

Cape Bojador, naming it Rio d'Oro in the mistaken belief that they had found the River of Gold. By 1444 they had passed the southern limits of the Sahara and reached Cape Verde, 'the green cape', on the frontier of 'the land of the blacks'. They called the local inhabitants there 'Guineas', after the Moroccan Berber word for 'blacks'.

They also began filing reports about elephants. The Portuguese captain Azurara sailed to the mouth of the Senegal River in 1450 and recorded that Africans enjoyed eating elephant meat but found no use for their tusks. The Venetian explorer Cadamosto, who travelled on Henry's ships to the Gambia River in 1456, developed a fascination about elephants and described eating elephant meat which had been boiled and roasted. 'To tell the truth,' he wrote, 'elephant meat is not very good. It seemed to me tough and unappetising and with little taste.' Cadamosto returned to Portugal with an elephant's foot, part of a trunk and a tusk which he duly presented to Henry.

Trade with 'Guinea', as the West African coast was known, soon burgeoned. It became sufficiently lucrative to attract a prominent Lisbon merchant, Fernão Gomes, in 1469 to acquire a five-year monopoly on trade beyond the Cape Verde Islands; in exchange he was required to pay an annual rent, commit his ships to explore 400 miles of new coastline each year, and sell to the Portuguese crown all the ivory he could obtain from local Africans.

Gomes's ships advanced rapidly around the great bulge of West Africa. In 1472, after dropping anchor off the estuary of the Pra River, his captains finally located the goldfields that had inspired Henry's expeditions, in an area that later became known as the Gold Coast. By agreement with local rulers, the Portuguese built a castle on the coast

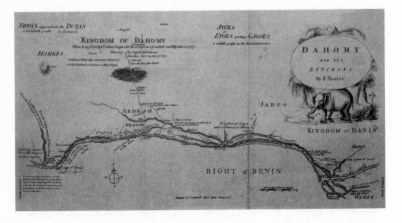

at a place they named El Mina, meaning 'the mine'. Within a few years they had secured an annual trade of some 20,000 ounces, a significant proportion of the world's supply.

Further to the east, in the dense rainforests of southern Nigeria, the Portuguese encountered the kingdom of Benin, where elephants and ivory played major roles in economic, political and cultural life. Ivory in Benin was a royal monopoly, belonging to the Oba, the hereditary king of the Edo people. One tusk of each elephant killed within the kingdom had to be presented to him and the other offered to him for purchase. The Oba, in turn, was a generous patron of the ivory carvers' guild, the Igbesanmwan. Ivory carving was an established art at his court, carried out with high skill. As well as producing regalia for the royal court, the Igbesanmwan turned out a variety of other ivory carvings for the wealthy elite—bowls, boxes, combs and bracelets, sometimes inlaid with copper or giltwork.

The Portuguese were impressed by the quality and duly commissioned work that they could take back to Europe: salt-cellars, forks, spoons and hunting horns. Designs for the salt-cellars usually depicted a group of standing

Portuguese grandees facing outwards or a set of equestrian figures accompanied by a naked angel. Similar work was commissioned from Temne-Bollom carvers in Sierra Leone.

The search for Prester John's empire meanwhile continued in earnest. Portuguese mariners crossed the equator in 1474; they reached the Congo estuary in 1483 and rounded the Cape of Good Hope in 1488.

In 1497 an expedition under the command of Vasco da Gama set sail from Lisbon carrying letters for various potentates he hoped to meet, including Prester John. Sailing up the east coast of Africa, da Gama's men reached the harbour of an offshore island called Mozambique where they were told by Arab traders that Prester John held many cities to the north. An officer of the fleet recorded: 'This information, and many other things which we heard, rendered us so happy that we cried with joy, and prayed to God to grant us health, so that we might behold what we so much desired.'

On reaching the Swahili port of Malindi, however, da Gama was diverted by an even greater prize. By chance he encountered there one of the most renowned Arab navigators of the time, Ahmad Ibn Majid, and persuaded him to show the Portuguese the sea route to India. Thus began an age of European maritime power in the Indian Ocean. Ibn Majid later regretted the help he had given to the Portuguese: 'Oh! Had I known the consequences that would come from them!' he wrote.

Following da Gama's return to Lisbon in 1499, the Portuguese sent out a series of armed expeditions to East Africa to enforce their control over its wealthy trading ports. Towns that refused to submit to Portuguese demands were bombarded, then pillaged. Zanzibar was the first to succumb in 1503; Mombasa was sacked in 1505; Kilwa, Mozambique Island and Sofala were also subjugated.

The quest to reach the land of Prester John was not forgotten. Portuguese envoys finally managed to establish contact with the Christian rulers of Ethiopia in their ancient mountainous kingdom. But little came of Portugal's hopes of forging a military alliance against their Muslim adversaries.

Nor did Portugal's commercial empire prosper as had been hoped. In 1506 a Portuguese commander reported that Sofala, the gateway to the Zimbabwe goldfields, was capable of supplying 4,000 tons of gold per year. Officials in Lisbon were jubilant at the news, believing that they had found another Gold Coast. A fortress was built at Sofala; trading posts were established along the Zambezi River valley; but only a comparative trickle of gold emerged.

There were, however, plenty of elephants. A Dominican missionary, João dos Santos, who arrived in Sofala in 1586 and later travelled up the Zambezi, reported: 'The number of elephants in this country is prodigious, so much so indeed that the inhabitants are obliged to pursue and make frequent hunting courses after them, to preserve from their ravage the lands they sow with rice and millet, in which lands these animals generally commit waste.'

The local Africans, he noted, were principally interested in elephants for meat, but there was also a considerable traffic in ivory exported to India.

He described the dangers of elephant hunting and recorded an incident in which two hunters had tracked down elephants they had wounded: 'One of these elephants had gone into a river, and with its trunk was throwing water over the other; this was lying on the bank, and in consequence the huntsman concluded it was dead. Approaching now somewhat nearer than was prudent, to the living one in the water, this elephant seized one of the two hunters with his trunk, and cast him with such violence on the body of the dead elephant as to deprive him of life, thus avenging the death of his comrade.'

---

Portuguese traders were soon followed into Africa by a host of other Europeans: the Dutch, English, French, Danes, Swedes and Brandenburgers, all seeking to gain from what was to become a relentless scramble for gold, for slaves and for ivory. Along the West African coast they built a string of fortified stations to protect their trade from European rivals. Rarely did they venture more than a few miles inland, relying on African middlemen for business.

As in the case of the Gold Coast, parts of the coastline became identified with particular commodities. Eastwards from the Gold Coast, along the Bight of Benin, where traffic in slaves was the main business, traders called it the Slave Coast. Westwards, along the surf-bound shore, where there were no good harbours and where dense forests lined the coast and the local population was thinly scattered, the main trade was in ivory; and here the coastline became known as the Tooth Coast or the Ivory Coast.

No Europeans were allowed to set up trading stations on the Ivory Coast, or to penetrate inland. The local population remained distrustful. Trade deals mostly took the form of 'smoke trading'. European ships, standing offshore, would fire a gun to attract attention. If local Africans had ivory or any other commodity to trade, they would send up smoke signals, then launch a canoe through the surf.

Elephants, however, were found throughout West Africa's coastal region, even populating islands fringing the shore; and ivory was obtainable from a string of trading posts. The Portuguese at Cape Verde were the first to put the ivory trade on a regular basis. Others followed intermittently. A Plymouth sea captain, William Hawkins, making the first English voyage to West Africa in 1530, picked up a cargo of 'Elephants' teeth' from Guinea before sailing on to Brazil. Ten years later John Landye, a captain in Hawkins' service, shipped 'one dozen elephants' teeth weighing one cwt'. In 1555 Thomas Windham, on a voyage to Guinea and Benin, brought back 250 tusks.

Hunting parties were sometimes organised. In January 1557 William Towerson recorded: 'This day wee tooke thirtie men with us and went to seeke Elephants, our men being all well armed and with harquebusses, pikes, long bowes, crossbowes, partizans, long swordes, and swordes and bucklers; wee found two Elephants, which wee stroke divers times with harquebusses and long bowes, but they went away from us and hurt one of our men.'

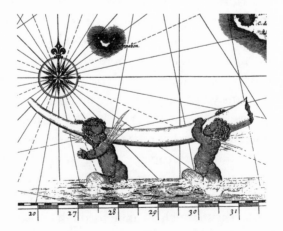

During the seventeenth century, after establishing shore bases in Upper Guinea, the English and Dutch exploited the ivory trade ruthlessly. In some areas there were reports that elephant herds had been slaughtered indiscriminately to meet the demand.

A Portuguese missionary, André de Faro, visiting Tarso Island in 1663, recorded his astonishment at the amount of ivory taken by an English ship:

Before my eyes, it loaded 28,000 teeth, many of which weighed four *arrobas* [58 kilos or 128 pounds], and there were numerous smaller ones. Every year, a ship comes to take a similar cargo. Judge from this how many elephants are killed here every year, because each one has only two teeth, and thus 14,000 elephants must be slain annually. This does not take account of the ivory that is purchased in the other rivers of Guinea, where there are similar factories, which dispatch other ships; and the Dutch are also buyers in the ports of these rivers. There are, therefore, more elephants in Guinea than there are cattle in the whole of Europe.

The missionary's arithmetic was clearly at fault, but his more general observation was accurate enough. By the mid-seventeenth century elephant herds on the coast of Guinea were noticeably depleted.

Further inland, they were still to be found in abundance. 'The inland countries of Benin ... Rio de Dalbary, Camerones, and several other adjacent countries, are so incredibly overcharged with these beasts that it is to be admired that the inhabitants live there,' wrote Willem Bosman, the chief Dutch factor at El Mina on the Gold Coast, in 1705. The young Scottish explorer Mungo Park, one of the first Europeans to venture into the interior of

West Africa, reported large herds inland during his journey
to the Niger River in 1795–7.

But even far inland the ivory trade eventually began to
take its toll. In the kingdom of Loango, north of the Congo
estuary, the Vili people were accustomed to using ivory
for personal ornamentation, for musical instruments, for
knives at table and for religious ritual. In the 1570s they
began to trade in ivory regularly with European trading
posts set up on the coast. By 1608 the Dutch alone were
buying twenty-three tons of ivory a year from Loango.
When their normal hunting grounds east of the coastal
plain could no longer meet the demand, the Vili ventured
further inland, sending caravans to markets as far inland as
Malebo Pool on the Congo River (or Stanley Pool as it later
became known). By the 1660s Vili traders, searching for
new sources of ivory, had to travel as far as Bukkameale, a

region known as 'The Mountains of Ivory'. It took them three months to get there and to return. And still the demand increased.

African traders were often puzzled by the European demand for ivory, as Mungo Park observed during his travels in Gambia:

Nothing creates a greater surprise among the Negroes on the sea coast, than the eagerness displayed by the European traders to procure elephants' teeth; it being exceedingly difficult to make them comprehend to what use it is applied. Although they are shown knives with ivory hafts, combs and toys of the same material, and are convinced that the ivory thus manufactured, was originally parts of a tooth, they are not satisfied. They suspect that this commodity is more frequently converted in Europe, to purposes of far greater importance; the true nature of which is studiously concealed from them, lest the price of ivory be enhanced. They cannot, they say, easily persuade themselves, that ships would be built, and voyages undertaken, to procure an article, which had no other value than that of furnishing handles to knives, etc. when pieces of wood would answer the purpose equally well.

Until the coming of Europeans, Africans hunted elephants mainly for meat. Elephant hunting was a dangerous profession. The only weapons available were spears, arrows, axes or a variety of trap devices. In Upper Guinea, Sape hunters hiding in trees plunged poisoned harpoons into passing elephants from above. Nalu hunters attacked with spears from close quarters. The Mandingas used poisoned spears thrown either from standing positions, or, occasionally, from horseback. Because of the danger, successful hunters were often celebrated in story and song.

Elephants were an important part of African culture. The Mende of Sierra Leone believed that their ancestors were elephants and that people became elephants when they died. Elephants were thus considered to be the physical form assumed by spirits of the dead. In many parts of Africa elephant imagery was used in rituals, ceremonies and masquerades. Social events such as weddings, funerals and initiations were conducted to the accompaniment of ivory side-blown trumpets, with sounds sometimes similar to the actual trumpeting of elephants.

Elephants were also adopted by African leaders as symbols of power and strength. In the Asante kingdom of the Gold Coast, the Golden Elephant Tail was the principal symbol of the kingdom's wealth; and elephant hide was used to support the Golden Stool, the most venerated symbol of all. Elephant ivory was usually considered the prerogative of chieftaincy, as in the case of Benin, with one tusk from each elephant kill being reserved for local rulers. Large unworked tusks were often displayed alongside a local ruler as a symbol of his authority as he sat in state at public events.

Ivory was also favoured for personal adornment, as bracelets, pendants, necklaces, armlets and leglets. It was turned into utensils and tools.

None of this activity, however, threatened the survival of elephant herds. Hunting elephants was principally carried out for subsistence purposes. But the growing demand for ivory from European traders turned hunting into a serious business enterprise.

In exchange for ivory, traders provided cloth, hardware, metals, beads and liquor. Then, from the mid-seventeenth century, they increasingly sold firearms. Firearms were in strong demand, both for slave raids and for hunting. By the 1690s flintlock muskets were in common use. They were inaccurate, tedious to load and cumbersome to operate.

Describing an incident at El Mina one morning in December 1700, when an elephant sauntered along the beach into a garden, Willem Bosman observed how difficult it was to kill with firearms:

Above one hundred shots were fired at him, which made him bleed to that degree . . . During all of which he did not stir, but only set up his ears . . .

But this sport was accompanied with a tragical event; for a Negro, fancying himself able to deal with him, went softly behind him, catched his tail in his hand, designing to cut a piece of it off; but the elephant being used to wear a tail, would not permit it to be shortened in his lifetime: wherefore, after giving the Negro a stroke with his snout, he drew him to him, and trod upon him two or three times; and if that was not sufficient, he bored in his body two holes with his teeth, large enough for a man's double fist to enter. Then he let him lie, without making any farther attempt on him; and stood still also whilst two Negroes fetched away the dead body, not offering to meddle with them in the least . . .

After remaining in the garden for about an hour, the elephant made his way to a river. 'Whilst the elephant stood there, the shooting began to be renewed, till at last he fell down; after which they immediately cut off his snout, which was so hard and tough that it cost the Negroes thirty strokes before they could separate it, which must be very painful to the elephant, since it made him roar . . . this elephant suffered above three hundred shots to be made at him, without any sign of being enraged or resistance . . .'

But the advent of firearms, coupled with European demand for ivory, was eventually to prove fatal for many of Africa's elephant populations. It was not long before the first extinctions occurred.

# 9
## *The Cape of Death*

When Jan van Riebeeck sailed into Table Bay in 1652 with instructions from the Dutch East India Company to establish a refreshment station there on the long haul between Europe and Asia, elephants were a common sight on the Cape peninsula. Company officials, under orders to make as much profit from the station as possible, hoped that sales of ivory would produce a useful source of income. Van Riebeeck's first mention of ivory in his journal is a record of three tusks bartered from local Khoikhoi hunters for nine ounces of tobacco.

The settlement at Cape Town and its surroundings steadily expanded, reaching forty miles inland by 1679 when Stellenbosch was founded. Year by year, the elephants' domain was reduced. During the eighteenth century nomadic stock farmers and hunters known as *trekboers* spread out from the Cape settlement, living with their families in ox-drawn wagons, always on the move. Ivory was a valuable part of their income. Some *trekboers* became professional ivory hunters.

The most successful elephant hunter of the time was Jacobus Botha, a *trekboer,* born in the 1690s, who fathered twelve sons and eventually retired to a farm in Swellendam with a fortune made from ivory. Recalling his exploits to an itinerant Swedish botanist, Carl Thunberg, in 1772, he boasted how he had often shot four or five in a day, sometimes twelve or thirteen and on two occasions, twenty-two elephants. In his old age Botha may have exaggerated his prowess as a hunter. But his fortune was real enough; and he readily remembered the time in his youth when elephants were so numerous in the Cape settlement that it was always easy to find them.

By 1760 elephants could no longer be found south of the Oliphants River on the western coast of the Cape colony; on the eastern coast the nearest elephant strongholds lay 500 miles distant from Cape Town. They remained plentiful in the eastern frontier region for a short while longer. A travelling Swedish scientist, Anders Sparrman, referred in 1775 to incredible numbers there. A government official in 1797 reported a herd 400 strong. But by 1830, when elephant hunting in the Cape colony was banned, only two small herds were left in the eastern Cape, one hidden deep in the Knysna forest and the other in the Addo bush country. Out of a Cape herd once estimated at 25,000, no more than a few hundred survived. It was the first mass extinction since Roman times.

Yet demand for ivory continued only to grow. As Europe and the United States entered an era of industrial revolution, bringing increased prosperity, there was a surge in demand for manufactured ivory products. Among the products popular with the burgeoning middle class were combs, cutlery handles and ornaments of every kind, all items that had found favour with wealthy elites down the centuries. But two new products brought about a massive increase in the use of ivory. One was piano keys; the other was billiard balls.

Britain, a leading market for ivory, imported an average of 66 tons a year between 1770 and 1800; during the 1820s its imports rose to an average of 190 tons a year; and during the 1830s the amount rose to 260 tons each year. Between the 1780s and the 1830s the price of ivory increased tenfold, with an impact that was felt around Africa.

In southern Africa the ivory hunters moved northwards, across the Orange River, venturing into new lands that were to become known as Bechuanaland and the Transvaal: 'a veldt virgin to rifle, and absolutely teeming with wild animal life', according to one description. They travelled with their wives, children, servants and livestock, established base camps for the season and hunted in all directions.

The weapons they used—long, single-barrelled muzzle-loaders called *roers*—were more advanced than the antiquated match-lock and wheel-lock muskets they had previously relied upon. But they were still cumbersome and heavy, weighing sixteen pounds and firing quarter-pound balls. While riding on horseback, hunters had to roll the ball in a linen patch, trim the corners, drop a torn powder cartridge down the barrel, ram the ball down with a loading rod and fit the percussion cap to the nipple. In order to kill an elephant with a single shot, they had to get within a range of fifty yards. The explosion and the recoil from muzzle-loaders were massive, sometimes knocking men out of the saddle.

The *trekboers* were joined by a new breed of sports hunter from England, who used ivory as the means to pay for their expeditions and to profit from them.

A British army officer, Captain William Cornwallis Harris was the first to embark on this new style of hunting safari. Setting out from Graaff-Reinet on the Cape frontier in 1836—'the last outpost of civilisation'—he had to wait until he reached the Magaliesberg hills, 500 miles further inland (near modern Pretoria) for his first sight of elephants.

Following a trail along the Sant River, he came to a rocky valley where 'a grand and magnificent panorama' opened before him: 'The whole face of the landscape was actually covered with wild elephants. There could not have been fewer than three hundred within the scope of our vision. Every height and green knoll was dotted over with groups of them, whilst the bottom of the glen exhibited a dense and sable living mass ... a picture at once soul-stirring and sublime.'

Harris lost no time in getting to work. Sending men ahead to drive the elephants up the valley towards him, he and his Khoikhoi hunters raked fire on the elephants as they charged by, killing and wounding many. Surveying the battlefield on the following day, Harris came across a surviving calf hovering about its dead mother, making mournful piping notes and trying to raise her with its trunk. Finally, it wrapped its trunk about Harris's leg and followed him back to the wagons. Moved by the incident, Harris had it fed and cared for, but it soon died.

Harris's account of his safari, first published in 1838, inspired others to follow suit.

Gordon Cumming, another former army officer, spent five years from 1844 on safari, gaining fame and fortune. The son of a Scottish baronet, with a huge physique and a vivid red beard, he hunted wearing a green and yellow Gordon kilt. Elephant hunting was his favourite sport. He found it 'overpoweringly exciting' and liked to ride close in, reloading his muzzle-loader from the saddle, and firing sometimes from as close as fifteen yards away. His hunts often turned into running battles lasting several hours, as he chased and fled his quarry across the bush.

Cumming wrote frankly about his exploits. He recorded one occasion when a large bull elephant he had wounded limped slowly to a nearby tree and stood there, watching him 'with a resigned and philosophic air'. 'I resolved to devote a short time to the contemplation of this noble elephant before I should lay him low.'

Cumming lit a fire, brewed some coffee and sipped it slowly.

Having refreshed myself, taking observations of the elephant's spasms and writhings between the sips, I resolved to make experiments on vulnerable points, and approaching very near, I fired several bullets at different parts of his enormous skull. He only acknowledged the shots by a salaam-like movement of his trunk, with the point of which he gently touched the wounds with a striking and peculiar action.

Surprised and shocked to find I was only prolonging the suffering of the noble beast, which bore his trials with such dignified composure, I resolved to finish the proceedings with all possible despatch, and accordingly opened fire on him from the left side. Aiming at the shoulder I fired six shots with the two-grooved rifle, which must have eventually proved mortal, after which I fired six shots at the same part with the Dutch six-pounder.

Large tears now trickled down from his eyes, which he slowly shut and opened; his colossal frame quivered convulsively. And falling on his side he expired.

As well as the hunt, Cumming enjoyed the bush carnival that followed an elephant kill. In his memoirs he vividly describes the moment when men arrived to carve up the carcass, stripping off layers of skin with assegais, chopping flesh away from the ribs and working from inside the carcass, covered in gore, to get at layers of fat found around the intestines. It was, he said, 'a scene of blood, noise and turmoil which baffles all description'.

Some of the elephant's skull bones were chopped into pieces and chewed raw. Much of the meat was cut into strips, up to twenty feet long and hung to dry as biltong for later use. Cumming's own preference was for baked elephant feet, carefully prepared in close-fitting pits, and sliced trunk.

At first, there were plenty of elephants to be found in these new lands. William Cotton Oswell, an English hunter who accompanied the missionary explorer David Livingstone on expeditions to Lake Ngami and the Zambezi River, teaching him how to survive in the African bush, described how, on reaching the Zouga River near Lake Ngami in northern Bechuanaland in 1848, they found a perfect paradise of elephants: 'I came, as I got clear of the bush, upon at least four hundred elephants standing drowsily in the shade of the detached clumps of mimosa trees. Such a sight I had never seen before, and never saw again. As far as the eye could reach, in a fairly open country, there was nothing but elephants.'

Livingstone noted that the local people had no use for ivory. 'We saw many instances of ivory rotting in the sun,' he wrote. 'The people called the tusks "marapo hela", "bones only", and they shared the fate of other bones.'

But following in the wake of such pioneering journeys came a host of hunters and traders. Livingstone returned to the Zouga a year later and recorded that within that time no fewer than 900 elephants had been slain on the Zouga River.

Some hunters and traders handed out weapons and ammunition to local Africans to kill whatever they could find. One Portuguese trader, João de Albasini, operating in 1845 in the South African lowveld, in an area which later became Kruger National Park, employed 400 black hunters to kill elephants and rhinos.

Gordon Cumming, for all his love of hunting, was not averse to trading guns for ivory to increase his fortune: 'Although I voted the trading an intense bore, it was nevertheless well worth a little time and inconvenience, on account of the enormous profit I should realise. The price I had paid for the muskets was £16 for each case containing twenty muskets; and the value of the ivory I required for each musket was upwards of £30, being about 3,000 per-cent, which I am informed is reckoned among mercantile men to be a very fair profit.'

The toll on elephant herds became ever greater in the 1860s when hunters were able to acquire new single-shot breech-loading guns to replace their old muzzle-loaders, enabling them to reload quickly.

By 1870, elephants in the Transvaal were virtually extinct. In an area that was soon to be known as South Africa, a population estimated at about 100,000 had all but been wiped out.

The hunters moved northwards again, crossing the Limpopo, venturing into Matabeleland and Mashonaland. One of the first to travel there, in 1864, was an English hunter, William Finaughty. He found that elephants in Matabeleland had yet to understand the meaning of gunfire. On one occasion, he killed six bull elephants standing in a river-bed, one after another, as none of them moved on hearing the sound of his shots.

In 1866 two legendary Boer hunters, Jan Viljoen and Petrus Jacobs, arrived in Matabeleland, accompanied by their wives, children and livestock. The following year they managed to kill 210 elephants, a record total for one season's hunting. An English hunter, Henry Hartley, who arrived in Matabeleland at the same time, subsequently spent most of his career hunting north of the Limpopo, ending with a tally of 1,200 elephants.

Hunting in Matabeleland and Mashonaland, however, involved new hazards. Much of the area was 'fly country', the domain of the tsetse fly, where horses could no longer be used. Rather than hunt elephants on foot, many hunters—such as William Finaughty and Henry Hartley—steered clear. Only the hardiest persevered.

For one young Englishman, Frederick Selous, this was just the kind of challenge he was seeking. After a journey of 1,000 miles from the coast, Selous arrived in Bulawayo, the Ndebele capital, determined to join the hunt for elephants.

His subsequent adventures were used by the novelist Rider Haggard when creating the character of Allan Quatermain, the great white hunter of *King Solomon's Mines* and other novels, whose daring exploits enthralled generations of English schoolboys.

Only twenty years old, Selous sought a meeting with the Ndebele king, Lobengula, to ask his permission to hunt. 'I said I had come to hunt elephants, upon which he burst out laughing, and said, "Was it not steinbucks [a small species of antelope] that you came to hunt? Why, you're only a boy!"'

After making further disparaging remarks about Selous' youthful appearance, Lobengula walked away. But Selous persisted and sought another meeting. 'This time he asked me whether I had ever seen an elephant, and upon my saying No, answered, "Oh, they will soon drive you out of the country, but you may go and see what you can do." '

Selous set out, full of youthful ambition, armed only with an old single-barrelled muzzle-loader, venturing deep into 'fly' country on foot. His chief companion was a Khoikhoi tracker named Cigar, a former Cape colony jockey whom William Finaughty had trained to hunt elephants on horseback.

Over the next three years, Selous killed seventy-eight elephants, all but one on foot. Returning to Bulawayo after his first season, he reported to Lobengula about his exploits. 'Why, you're a man!' retorted Lobengula. 'When are you going to take a wife?'

Selous developed great respect for the intelligence, resourcefulness and stamina of elephants, admiring their ability to avoid detection, to move stealthily and to climb steep hillsides. Describing the precipitous hills and deep narrow ravines of the Zambezi escarpment, he recorded: 'At first sight, many of these cliffs appeared inaccessible to any animal but a baboon; but we found that the elephants had made regular paths up and down many of them, which paths zigzagged backwards and forwards like a road down a Swiss mountain, and in some places great blocks of stone had been forced aside by the efforts of these bulky engineers in order to render their footing more secure.'

Selous had many narrow escapes. Once on horseback, having searched for elephants all day, he came across a herd between two rivers in Mashonaland and, though his horse was exhausted, he decided to pursue them. After killing three elephants, he went for a fourth. 'The fourth I tackled cost me six bullets and gave me a smart chase, for my horse was now dead beat. I only got away at all by the skin of my teeth as, although the infuriated animal whilst charging trumpeted all the time like a railway engine, I could not get my tired horse out of a canter until he was close upon me, and I firmly believe that had he not been so badly wounded he would have caught me. I know the shrill screaming sounded unpleasantly near.'

Despite his narrow escape, he decided to pick out yet one more. He gave the elephant a shot behind the shoulder, cantered up to within thirty yards and fired a second shot. He was just reloading when the elephant charged.

Digging the spurs into my horse's ribs, I did my best to get him away, but he was so thoroughly done that, instead of springing forward, which was what the emergency required, he only started at a walk and was just breaking into a canter when the elephant was upon us. I heard two short sharp screams above my head, and had just time to think it was all over with me, when, horse and all, I was dashed to the ground. For a few seconds, I was half-stunned by the violence of the shock, and the first thing I became aware of was a very strong smell of elephant.

The elephant was on its knees, its head and tusks on the ground, with Selous pressed down under its chest. 'Dragging myself from under her, I regained my feet and made a hasty retreat, having had rather more than enough of elephants for the time being.'

In his first years as an elephant hunter, Selous made a good living from ivory. But even along the Zambezi River,

elephants eventually became scarce. Writing to his father in
1877, Selous remarked: 'On this side of the river elephant
hunting is at an end, all the elephants being either killed or
driven away.' He returned to England in 1881, seeing no
future in the business. 'I had already spent ten years of my
life elephant hunting in the interior and every year elephants were becoming scarcer and wilder south of the
Zambezi, so that it had become almost impossible to make
a living by hunting at all.'

Elsewhere in southern Africa the same results occurred.
By the 1880s the era of elephant hunting there was virtually
over. The great herds had vanished.

'The story of the elephant in South Africa [southern Africa] may be described as one long tragedy of extermination,' wrote H.A. Bryden in the *Fortnightly Review* in 1903. 'Such has been the rate of extermination that the wild elephant has now practically ceased to exist south of the Cunene and Zambezi Rivers.'

Further north, a similar slaughter was under way.

# 10
## *Zanzibar's Tune*

Lying twenty miles from the mainland, Zanzibar, in the nineteenth century, was the greatest commercial centre on the East African coast, a meeting place for slave-traders, ivory dealers and spice merchants and a base from which European explorers could venture into the vast uncharted territories of the African interior. Its anchorage was crowded with Arab dhows and square-rigged merchantmen, Americans from Salem, Spaniards from Cuba, French slavers from the Mascarene Islands, Indiaman from Bombay. From his palace on the island, the Sultan of Zanzibar claimed authority over trade routes stretching far inland, as far as the great lakes of central Africa. 'When they pipe in Zanzibar,' it was said, 'people dance on the shores of the great lakes.' Throughout the nineteenth century the tune that Zanzibar called was for ivory and slaves, in ever growing numbers.

Since gaining control of the region from the Portuguese in the seventeenth century, Omani Arabs from the Persian Gulf had turned Zanzibar from a small colonial outpost into a thriving entrepôt dominating trade along the coast and other offshore islands. So attractive was Zanzibar as a base, compared to the barren coast lands of the Persian Gulf, that Sultan Seyyid Said, after visiting the island in 1828, decided to forsake Muscat and to transfer his government to the island.

Abounding with ambition for his new capital, Said laid the foundations for a prosperous new state. Using slave labour, he promoted the development of clove plantations which eventually made Zanzibar a leading world producer; he signed commercial treaties with the United States, Britain and France; and he set his sights on creating a new commercial empire in the African interior.

Hitherto, trade with the mainland had been confined largely to a narrow coastal strip. Beyond the coast lay the *nyika*, a barrier of arid thorn scrub that was thinly populated. It was not until the early nineteenth century that African traders from Nyamweziland in central Tanzania pioneered new routes to the coast, a journey taking three months, bringing ivory and slaves they knew were in demand from Arab merchants there. In 1811 a British naval officer, Captain Thomas Smee, reported from Zanzibar that Nyamweziland abounded in elephants.

As the demand for both ivory and slaves increased, Arab dealers moved inland, seeking to control the trade and setting up prosperous settlements of their own deep in the interior. By the 1830s they had penetrated to the shores of Lake Tanganyika, 1,000 miles inland, and reported finding lands of potentially fabulous riches where ivory was used for making doorposts and fencing pigsties.

The main trading centre in the interior became an Arab settlement at Tabora near the Nyamwezi capital at Unyanyembe. The Arab community there lived in considerable comfort. Their houses were furnished with Persian carpets and luxurious bedding. They maintained extensive gardens with orchards and pastures for livestock. Fine foods were imported. And slaves and concubines attended their needs.

From Tabora, hunting parties armed with muzzle-loaders and heavy spears spread out across Nyamweziland searching for elephants and slaves. The Nyamwezi participated not only as hunters but as porters. Ivory caravans sent down to the coast consisted of hundreds, sometimes thousands of porters. Most were hired for the round trip, to carry back merchandise; some were slaves to be sold. In 1848 the Nyamwezi sent a caravan 2,000-strong to the coast with gifts for the sultan.

Beyond Tabora, the caravan routes branched out in all directions further into the interior: to the northwest around Lake Victoria into Uganda; to the southwest around the southern end of Lake Tanganyika to Katanga; and directly westwards to the lake where another thriving Arab settlement was established at Ujiji. From Ujiji's small harbour, dhows crossed Lake Tanganyika, taking traders to the fabled elephant country of Manyema in the eastern Congo.

It was along 'the ivory road' that two English explorers, Richard Burton and John Speke, set out in 1857 on their journey from Zanzibar to the great lakes, hoping to solve the enduring mystery of the source of the Nile. Burton described the caravans he encountered, winding their way 'like a monstrous land-serpent' over the plains.

Flying the plain, blood-red flag of a Zanzibar expedition, they were led by a guide, a *kiongozi*, dressed in a bright red gown and wearing a headdress of a black and white colobus monkey skin. Behind the *kiongozi* came the ivory porters, 'their shoulders often raw with the weight'. Two men were needed to carry the heavier tusks which were tied to a pole, with cowbells attached to their points sounding out as the caravan moved on. The ivory porters were followed by cloth-bearers. 'Behind the cloth-bearers struggles a long line of porters and slaves, laden with the lighter stuff, rhinoceros teeth, hides, tobacco, brass-wire . . . In separate parties march the armed slaves . . . the women, and the little toddling children, who rarely fail to carry something, be it only a pound weight.'

After 134 days on the march, Burton and Speke completed the 600-mile journey to Tabora. It took them another sixty days to reach Ujiji. During their return, Speke complained about the poor opportunities for hunting. 'This is a shocking country for sport; there appears to be literally nothing but elephants, and they, from constant hunting, are driven from the highways.'

In his account, Burton included a description of local elephant hunting. It was, he said, a solemn and serious undertaking. Before departing, hunters joined in dancing, singing and drinking and fortified themselves with traditional potions. Once a herd had been detected, their objective was to isolate one tusker from the others; then they formed a circle around it, hurling spears.

The baited beast rarely breaks . . . through the frail circle of assailants: its proverbial obstinacy is excited; it charges one man, who slips away, when another, with a scream, thrusts the long stiff spear into its hindquarters, which makes it change intention and turn fiercely from the fugitive to the fresh assailant. This continues till the elephant, losing breath and heart, attempts to escape; its enemies then redouble their efforts, and at length the huge prey, overpowered by pain and loss of blood trickling from a hundred gashes, bites the dust.

Other travellers later made use of the same ivory road. When Henry Stanley, a journalist on the staff of the *New York Herald,* was commissioned by its proprietor to find David Livingstone, who had disappeared into the African interior five years previously on an expedition to find the source of the Nile, his starting point was the ivory road to Ujiji. Crossing to the mainland from Zanzibar, Stanley marched inland at the head of a caravan which included porters, armed guards, cooks, a guide, an interpreter, two

British sailors and a dog named Omar. Eight months later, in November 1871, he reached Ujiji where, by chance, Livingstone had arrived from Manyema only a week before.

Since setting out from Zanzibar in 1866, Livingstone had wandered about central Africa searching for the source of a river where it did not exist, but believing in the end that he had found it. Though he was appalled by the depredations of the slave trade, he spent most of his time in the company of Arab traders who profited from it, depending on them for food, shelter and medicine and for nursing him through illness. In 1869, after recovering in Ujiji from a severe bout of pneumonia, he had decided to accompany an Arab merchant who was leading an ivory expedition to Manyema, expecting the journey to last for no more than a few months. But he did not return for two years.

Livingstone found Manyema thick with elephants. Collecting ivory there, he said, was like 'gold-digging', so much of it lay on the ground. But the area was also swarming with Arab traders, plundering at will. They employed armed bands not only to hunt elephants but to extract ivory from the local population. At any sign of resistance, villagers were murdered, their houses looted and burned. Livingstone himself witnessed a massacre in the market at Nyangwe, a Zanzibari settlement on the Lualaba River. Horrified by what he had seen, he abandoned plans to explore further westwards and returned to Ujiji, sick, exhausted and destitute, and only too relieved when Stanley arrived a week later.

In the 1870s a new empire arose in Manyema, based on ivory and slaves. It was established by a powerful Zanzibari trader, Hamed bin Muhammad el Murjebi, better known as Tippu Tip, a nickname derived from his habit of nervously blinking his eyelids. Ostensibly, Tippu Tip owed allegiance to the Sultan of Zanzibar, but in reality he acted as an independent ruler with control over much of the eastern Congo.

Both his grandfather and his father had taken part in the caravan trade to Lake Tanganyika, and Tippu Tip's own earliest journeys were with Nyamwezi caravans travelling round the south end of Lake Tanganyika to Katanga. He was also active in his youth in slave-raiding, as he recalled in his memoirs: 'I went into every part of Zaramu country and in the space of five days had seized 800 men. They called me Kingugwa Chui [the leopard], because the leopard attacks indiscriminately, here and there. I yoked the whole lot of them together and went back with them to Mkamba.'

On his travels, Tippu Tip played host to several European explorers. In 1867, when David Livingstone was stranded in northern Zambia, weak with fever and hunger, Tippu Tip had supplied him with provisions, given him a letter of introduction to a neighbouring African king and assigned guides to accompany him on his way. He had helped John Speke on his second expedition to discover the source of the Nile in the 1860s. He had also been hospitable to Captain Cameron, a British naval officer who in the 1870s became the first European to cross Africa from the east coast to the west.

The empire that Tippu Tip established in the 1870s extended over an area of some 250,000 square miles. From his capital, Kasongo, on the Lualaba River, he appointed officials, collected tributes, built roads, laid out plantations and imposed control over the trade in ivory and the hunting of elephants. His raiding parties were feared throughout Manyema.

In 1876, Henry Stanley, on his second African expedition, travelled through Manyema on his way from Ujiji to the Lualaba River, noting how the country had been ravaged by slave-traders. On arriving in Kasongo, Stanley nevertheless struck up a cordial relationship with Tippu Tip and sought his help. Stanley's objective was to test Livingstone's theory that the Lualaba flowed northwards into the Nile by following it downstream, wherever it led, until he reached the mouth. He had brought with him a forty-foot boat named *Lady Alice* in honour of a seventeen-year-old American heiress with whom he had fallen in love. Built in five sections for porterage, *Lady Alice* had already proved its worth on Stanley's expedition during his circumnavigation of Lake Victoria and Lake Tanganyika. But the journey down the Lualaba involved new hazards: it would lead him into the depths of the rainforest into completely unknown territory where cannibal tribesmen were reputed to live, and Stanley feared that his porters would desert.

Not even Tippu Tip had ventured there. Nor did he see any reason why he should. As he told Stanley:

If you Wazungu [white men] are desirous of throwing away your lives, it is no reason we Arabs should. We travel little by little to get ivory and slaves, and are years about it—it is now nine years since I left Zanzibar—but you white men only look for rivers and lakes and mountains, and you spend your lives for no reason, and to no purpose. Look at that old man who died in Bisa [David Livingstone]! What did he seek year after year, until he became so old that he could not travel? He had no money, for he never gave us anything; he bought no ivory or slaves; yet he travelled further than any of us, and for what?

Yet Tippu Tip recognised that the journey would give him an opportunity of extending his empire of slaves and ivory. The rainforest was said to harbour large herds of elephant and nobody who lived there knew what tusks were worth. He was also impressed when Stanley demonstrated his advanced weaponry—a repeating rifle capable of firing fifteen rounds. And when Stanley offered to pay the sum of 5,000 Maria Theresa dollars, he soon agreed to accompany him with an armed party, though for no longer than sixty days.

So, in November 1876, they set off, entering 'the dreaded black and chill forest', as Stanley described it, 'bidding farewell to sunshine and brightness'. Day after day, they persevered, descending through the twilight, hacking their way through riverside jungle, fighting off hostile tribesmen, and braving malaria, dysentery and smallpox.

After travelling together for 250 miles down the Lualaba, Stanley and Tippu Tip parted company. Stanley continued his perilous journey downstream for another 1,500 miles, eventually reaching the mouth of the Congo River on the Atlantic coast seven months later with the remnant of his party, starving, haggard and close to defeat.

Tippu Tip went off to collect another fortune in ivory from virgin territory. 'At this time the locals did not use ivory as exchange,' he recalled. 'They hunted elephant and ate the meat but used the tusks in their homes for a stockade. With others they made pestles and mortars for their cooking bananas.' After a month he returned to Kasongo with his booty, having opened up a new domain to plunder.

In 1882, after spending twelve years in the interior, Tippu Tip decided to visit Zanzibar, setting out from Manyema with a huge caravan of ivory. As the caravan passed through Mpwapwa, 200 miles from the coast, its progress was observed by a British mariner, Alfred Swann.

Swann had been hired by the London Missionary Society to transport a boat from the coast to Ujiji, to reassemble it there and then to sail it on Lake Tanganyika on missionary business. He had arrived in Zanzibar fired with missionary zeal to destroy the slave trade but was shocked to find that his own porters were themselves slaves. He was even more horrified by what he saw of Tippu Tip's caravan.

As they filed past, we noticed many chained by the neck. Others had their necks fastened at the forks of poles about six feet long, the ends of which were supported by the men who preceded them. The women, who were as numerous as the men, carried babies on their backs in addition to a tusk of ivory or other burden on their heads. They looked at us with suspicion and fear, having been told, as we subsequently ascertained, that white men always desired to release slaves in order to eat their flesh, like the Upper Congo cannibals.

It is difficult adequately to describe the filthy state of their bodies; in many instances, not only scarred by the cut of a 'chikote' [a raw-hide whip] . . . but feet and shoulders were a mass of open sores, made more painful by the swarms of flies which followed the march and lived on the flowing blood. They presented a moving picture of utter misery, and one could not help wondering how any of them had survived the long tramp from the Upper Congo, at least 1,000 miles distant . . .

The headmen in charge were most polite to us, as they passed our camp . . . Addressing one, I pointed out that many of the slaves were unfit to carry loads.

To this he smilingly replied: 'They have no choice! They must go, or die!' . . .

'Have you lost many on the road?'

'Yes! Numbers have died of hunger!'

'way?'

are too well guarded. Only those who become pos-

the devil try to escape; there is nowhere they could

y should go.'

you do when they become too ill to travel?'

em at once! . . . For if we did not, others would

y were ill in order to avoid carrying their loads. No!

eave them alive on the road; they all know our
custom.'

'I see women carrying not only a child on their backs, but, in
addition, a tusk of ivory or other burden on their heads. What
do you do in their case when they become too weak to carry
both child and ivory? Who carries the ivory?'

'She does! We cannot leave valuable ivory on the road. We
spear the child and make her burden lighter. Ivory first, child
afterwards!'

Swann raged: 'Ivory! Always ivory! What a curse the elephant has been to Africans. By himself the slave did not pay
to transport but plus ivory he was a paying game.'

The trade was indeed profitable. Stanley calculated that a
pound of ivory costing one cent in Manyema was worth 110
cents in Tabora and 200 cents in Zanzibar. Zanzibar
became the richest seaport in tropical Africa from its trade
in ivory, slaves and cloves. But it was always ivory that was
its most important export. It was far more lucrative than
the trade in slaves. By 1890, Zanzibar was supplying three-
quarters of the world's trade in ivory.

But the cost of all this to East Africa's elephant populations was massive. According to estimates based on trading

and auction records, some 60,000 elephants were killed each year. The great herds were steadily decimated.

After leading a Royal Geographical Expedition to the great lakes in 1879–80, Joseph Thomson reported:

People talk as if the ivory of Africa were inexhaustible. It is commonly supposed that, if European traders could but establish themselves in the interior, fortunes could be made. Nothing could be more absurd. Let me simply mention a fact. In my sojourn of fourteen months, during which I passed over an immense area of the Great Lakes region, I never once saw a single elephant. Twenty years ago they roamed over those countries unmolested, and now they have been almost utterly exterminated. Less than ten years ago Livingstone spoke about the abundance of elephants at the south end of Tanganyika—how they came about his camp or entered the villages with impunity. Not one is now to be found. The ruthless work of destruction has gone on with frightful rapidity.

Other regions of eastern Africa faced the same threat.

# II
## *The Khartoum Run*

In 1839 an expedition of ten boats commanded by a Turkish sea captain set sail from Khartoum heading for the upper reaches of the White Nile on orders from Egypt's ruler, Muhammed Ali, who hoped that gold and other riches would be found in the vast unexplored regions of the southern Sudan. Barring their way was the sudd, a vast swamp of papyrus and rotting vegetation which no outsiders had previously managed to penetrate. After two attempts, the expedition succeeded in breaking through the sudd and reached Gondokoro, a village 1,000 miles south of Khartoum, but found no trace of gold. Instead, it opened a route into one of the greatest reserves of elephant country in Africa stretching for hundreds of thousands of square miles along the White Nile and its tributaries.

News of this new highway into the interior drew an increasing number of merchants, missionaries, ivory traders and adventurers to Khartoum, the flyblown garrison town that Muhammed Ali had established in the 1820s on a promontory of land formed by the confluence of the White Nile and the Blue Nile, intending to extend the boundaries of his empire. Some came from Europe—Greeks, Italians, Austrians, French and English—founding a cosmopolitan community with its own comfortable houses, shops and churches. A monthly camel post kept them in touch with the outside world, and luxuries like wine, Bass's pale ale,

French biscuits, soaps and perfumes were imported via the northern desert.

One of the first travellers to explore southwards was a Welsh mining engineer, John Petherwick, who had previously been employed by Muhammed Ali to search for coal deposits in western Sudan, a venture that was unsuccessful. Appointed a British vice-consul in 1850, he made several expeditions to the Bahr el Ghazal, one of the main tributaries of the White Nile, returning with a hoard of ivory. The only use that local Zande tribesmen made of ivory, he reported, was for ornaments such as bracelets and necklaces. Ivory could readily be obtained in exchange for beads, cowries or copper bracelets.

An ivory 'rush' was soon under way. Each year, in November, when the north winds began to blow at Khartoum, a flotilla of trading boats set out up the White Nile on an annual expedition to collect ivory. In 1851 there were a dozen boats; by the end of the season they had collected some 400 quintals of ivory, about 18,000 kilos (40,000 pounds), costing them about 1,000 francs in beads. Sold in Cairo, it was worth 100,000 francs. In 1856 more than forty boats set out, returning with 1,400 quintals. 'I believe the annual imports will increase,' reported Petherwick, 'the traffic being now better understood and more vigorously prosecuted than before.'

Indeed, the trade became increasingly rapacious. Once supplies of elephant and ivory close to the Nile were exhausted, traders mounted expeditions inland employing armed gangs of Arab hirelings to establish fortified camps, *zeribas,* from where they sent out raiding parties. As well as plundering for ivory, they traded in slaves, taking advantage of local tribal rivalries to encourage villagers to attack their neighbours, abduct women and children and drive off herds of cattle and sheep which were then ransomed for more ivory. A vast swathe of the southern Sudan became known as *zeriba* country.

One of the pioneers of this *zeriba* trade was a Frenchman, André de Balzac, who became known as the King of the White Nile. After making a reconnaisance of Dinka territory in 1854, he established a *zeriba* eight days' march into the interior. According to missionaries, he adorned his stockade with the heads of his victims and created such terror that whole tribes fled the neighbourhood. His ivory business was so successful that after his first season he needed 500 porters to transport his ivory to the banks of the Nile.

By 1862 the number of boats setting out from Khartoum on the annual expedition had reached some 120. They carried parties of up to 300 armed Arab guards, many of them former criminals, hired by traders to act as their private armies on raids in the south. 'There are no longer merchants but only robbers and slavers on the White Nile,' the Austrian consul reported from Khartoum.

The profits from these expeditions were considerable. An English traveller, Samuel Baker, who visited Khartoum in 1862, calculated that, in a good season, a trader employing a party of 150 men could obtain about 9,000 kilos (20,000 pounds) of ivory, valued in Khartoum at about £4,000. He usually paid off the men in slaves and cotton pieces. This still left him with a surplus of 400 or 500 slaves which he could sell for £5 or £6 each.

Baker, a wealthy big-game hunter, regarded as one of the finest shots in England, had arrived in Khartoum after spending a year wandering in the eastern Sudan accompanied by

his Hungarian lover, Florence Ninian von Sass, whom he had bought on sight for £7 at a Turkish slave market. They had become inseparable companions. During elephant hunts on horseback, when Baker dismounted to pursue a quarry on foot, Florence would hold his bridle, galloping to his side as soon as she heard a shot, ready for him to remount in case pursuit was needed.

It was while exploring the tributaries of the Blue Nile in eastern Sudan that Baker first encountered Baqqara Arabs, cattle-owning nomads with a long tradition of duelling with elephants. Armed only with sharp lances, they hunted elephants on horseback, usually in pairs, with one rider galloping in front of the elephant to distract its attention, while the other rode close in from behind, dismounting at full gallop and thrusting his lance into the elephant's hindquarters. The elephant usually turned on the attacker. While he tried to escape either on horse or on foot, the other rider moved in, dismounted and made a similar attack. 'In this dangerous hand-to-hand fighting,' Baker remarked, 'the hunter is often the victim.'

A previous traveller, James Bruce, making his way down the Blue Nile in 1772, witnessed the Baqqara hunting elephant, but with the rider and a lancer astride the same horse. 'As soon as the elephant is found feeding, the horseman rides before him as near his face as possible; or, if he flies, crosses him in all directions, crying out, "I am such a man and such a man; this is my horse, that has such a name; I killed your father in such a place, and your grandfather in such another place, and I am now come to kill you; you are but an ass in comparison to them." This nonsense he verily believes the elephant understands.'

Instead of running for safety, the elephant chased the horse. 'After having made him run once or twice in pursuit of the horse, the horseman rides close up alongside of him, and drops his companion just behind on the off side; and while he engages the elephant's attention upon the horse, the footman behind gives him a drawn stroke just above the heel . . .

'This is the critical moment; the horseman immediately wheels round, and takes his companion up behind him, and rides off full speed after the rest of the herd.'

Bruce concluded, as Baker did, that it was a dangerous profession: 'Dextrous . . . as the riders are, the elephant sometimes reaches them with his trunk, with which he dashes the horse against the ground, and then sets his feet upon him, till he tears him limb from limb with his proboscis; a great many hunters die this way.'

During his time in the Sudan, Baker took copious notes of local hunting methods. The most common method of hunting elephant, he recorded, as Pliny had done 2,000 years before, was the pitfall, often dug on the approaches to rivers.

The night arrives, and the unsuspecting elephants, having travelled many miles of thirst wilderness, hurry down the incline towards the welcome river. Crash goes a leading elephant into a well-concealed pitfall! To the right and left the frightened members of the herd rush at the unlooked-for accident, but there are many other pitfalls cunningly arranged to

meet this sudden panic, and several more casualties may arise, which add to the captures on the following morning, when the trappers arrive to examine the position of their pits. The elephants are then attacked with spears.

Another common method of hunting elephant was the use of fire. In the dry season, when the tall grass burned easily, as many as 1,000 men would form a line several miles long around an elephant herd and trap it in a ring of fire.

The men advance with the fire, which rages to the height of twenty or thirty feet. At length the elephants, alarmed by the volume of smoke and the roaring of the flames, mingled with the shouts of the hunters, attempt an escape. They are hemmed in on every side—wherever they rush, they are met by an impassable barrier of flames and smoke, so stifling, that they are forced to retreat.

Meanwhile the fatal circle is decreasing; buffaloes and antelopes, likewise doomed to a horrible fate, crowd panic-stricken to the centre of the encircled ring, and the raging fire sweeps over all. Burnt and blinded by fire and smoke, the animals are now attacked by the savage crowd of hunters, excited by the helplessness of the unfortunate elephants.

After spending six months in Khartoum, Baker embarked with Florence on an expedition up the Nile in December 1862, hoping to find its source. He had also been asked by the Royal Geographical Society to look out for two explorers, John Speke and James Grant, who, after setting out from Zanzibar on their own search for the source of the Nile, had been missing for a year.

In March 1863 Baker's expedition reached Gondokoro, the main settlement in the south, which he described as 'a perfect hell' where traders and their armed hirelings were forever drinking, quarrelling and firing their guns wildly in the air.

By chance, only two weeks later, Speke and Grant staggered into Gondokoro, returning down the Nile after discovering its main source at Ripon Falls on the northern shore of Lake Victoria. But they encouraged Baker to continue his journey and explore a possible second source, a lake named Lŭta Nzigé lying northwest of Lake Victoria, which they had not managed to reach.

For the next two years, Baker and Florence wandered about the upper reaches of the Nile, beleaguered by incessant local wars, braving constant danger, often ill with fever, short of food and supplies and dependent for survival on slave- and ivory-traders. But they succeeded in finding Lŭta Nzigé, renaming it Lake Albert in honour of Queen Victoria's late consort, and around its edges they encountered wonderful elephant country.

Travelling downstream from the lake, Baker recorded seeing moving parallel to his caravan 'a vast concourse of elephants, grouped in parties of varying size from ten to one hundred animals while single bulls dotted the landscape, with their majestic forms in all directions'. The procession of elephants, wrote Baker, stretched for two miles.

Baker returned to the Sudan in 1869, hired by the Khedive of Egypt to act as governor-general of Equatoria, its vast, ill-defined southern region, and to impose some sort of order on it. It was an impossible task, as Baker realised shortly after arriving in Khartoum, when he was informed that the government had rented out the entire White Nile to traders. 'I was to annex a country that was already leased out by the Government. My task was to suppress the slave trade, when the Khartoum authorities well knew that their tenants were slave-hunters; to establish legitimate commerce when the monopoly of trade had already been leased to traders; and to build a government upon sound and just principles, that must of necessity ruin the slave-hunting and ivory-collecting parties of Khartoum.'

The next governor-general of Equatoria, Colonel Charles Gordon, a British army officer, on taking up his appointment in 1874, declared all ivory to be a government monopoly and met with some success. But what finally brought the Khartoum run to a halt was the Mahdi's uprising in the 1880s which swept the Khedive's regime out of the Sudan.

In the south, Equatoria province, cut off by the rebellion, was left in the hands of its governor-general Edouard Schnitzer, a German medical doctor and naturalist, better known as Emin Pasha. Emin continued to trade in ivory long after links down the Nile had been severed, accumulating a vast stockpile in government warehouses. It was his fate and the fate of his ivory hoard that were to inspire yet another European expedition.

# 12
## Heart of Darkness

---

The vast expanses of the Congo remained impenetrable to explorers and traders alike for longer than any other region. In the west, about fifty miles inland from the Atlantic coast, on the north shore of the Congo River, there was a trading post at Boma, where a small band of Europeans lived, rough hard-bitten men, used to wielding the whip and the gun, who never ventured further inland than the cataracts and canyons lying upriver. In the east, 1,000 miles away, Tippu Tip presided over his central African empire, trading in ivory and slaves. What lay in between was unknown.

Then, in 1877, Stanley staggered into Boma at the end of his epic journey across Africa, having found his way down the Congo River. It had taken him five months to cover the last 180 miles as he struggled desperately through a series of thirty-two rapids, losing men and boats. It was, he wrote, 'the wildest stretch of river that I have ever seen'.

Stanley's expedition unlocked the entire Congo region. For it showed that beyond the cataracts and canyons which had hitherto blocked exploration inland lay a web of interconnecting rivers, navigable by steamboat, running for thousands of miles into the interior. In Europe, Stanley's achievement in 'filling in Africa's blank space' was acclaimed the greatest feat of exploration of the century.

On his return to London, Stanley campaigned vigorously for European powers to open up the Congo to 'trade and civilisation'. The British government, which he approached first, declined to take an interest. But in Brussels, the Belgian king, Leopold II, leaped at the opportunity. An ambitious, greedy monarch, forever scheming, Leopold had long dreamed of establishing colonies abroad and enriching himself.

In 1879 Leopold hired Stanley to build a private empire for him in the Congo and exploit whatever wealth he could find. Ivory was his main hope. 'I am desirous to see you purchase all the ivory which is to be found in the Congo,' he wrote to Stanley.

Stanley worked tirelessly on Leopold's behalf. He spent two years building a wagon road inland from the first cataract at Yellalla Falls, over the Crystal Mountains, to Malebo Pool, a name that Stanley changed to Stanley Pool. On the south side of Stanley Pool, he established his main station near a village called Kinshasa which eventually became Leopoldville. From there, he launched a fleet of steamboats on the river which ventured further upstream, establishing new stations ever deeper into the interior. The furthest outpost was at Stanley Falls, 1,000 miles upriver from Stanley Pool, which marked the upper limit of navigation on the main stretch of the Congo River. Stanley had passed by the falls on his way to the Atlantic in 1877. He travelled there again in 1883 and found the entire region ravaged by slave-traders from Manyema who had followed in his wake.

By the time Stanley had completed his contract in 1884, he had signed treaties with some 400 African chiefs, persuading them to give up their sovereignty, and laid the foundations for Leopold's personal empire. With the approval of other European powers, Leopold declared himself 'King-Sovereign' of the Congo Free State in 1885. It was an area of 1 million square miles, seventy-five times the size of Belgium, bigger than England, France, Germany, Spain and Italy combined, one-thirteenth of the African continent, and it was all owned by Leopold. Leopold's principal objective henceforth was to amass as large a fortune as possible from this new estate.

Everything depended on how much ivory the network of river stations could deliver. They were no more than pinpoints in the forest, a few buildings with thatched roofs and shady verandahs, sheltered by palm trees, with the flag of the Congo Free State—a blue standard with a single gold star—flying from a pole. But they provided the backbone on which Leopold's empire was constructed. From there, his agents launched hunting expeditions and ivory raids, acquiring tusks in exchange for a few beads or brass rods or simply by confiscating them. Local inhabitants were prohibited from selling or delivering ivory to anyone else; nor were they allowed to receive payment in money. The agents meanwhile were paid on a commission basis: the more ivory they collected, the more they earned. Their methods of obtaining ivory and of conscripting porters to carry it consequently became increasingly ruthless.

As well as functioning as ivory-collection points, Leopold's river stations served as military outposts. His rule ultimately came to depend on the Force Publique, an army composed of white officers and African auxiliaries, notorious for brutal conduct, which eventually consumed half of the state's budget.

The Congo Free State was constantly afflicted by revolts, mutinies and uprisings. In 1886 Arab traders and their Manyema auxiliaries overran the river station at Stanley

Falls. So desperate was Leopold to regain control that he offered Tippu Tip, the slave-trader, the post of governor-general of the eastern Congo, with a free hand to exploit ivory and whatever other riches he could find. Tippu Tip duly accumulated another fortune in ivory. 'Life was very good in Stanley Falls,' he recalled. 'Trade was wonderful, and the number of tusks coming in was staggering.'

In 1890 a thirty-two-year-old Polish seaman named Konrad Korzeniowski arrived in the Congo to work as a river-boat captain. Eight years later, having adopted the name of Joseph Conrad, he used his experiences to write a novel exposing the madness of greed and corruption that overtook Leopold's Congo Free State in those years. Titled *The Heart of Darkness*, it became one of the most enduring novels of modern times.

The narrator of *The Heart of Darkness*, Charlie Marlow, is hired by an ivory-trading company to sail a steamboat up an unnamed river. His destination is a trading post called the Inner Station run by one of the company's most outstanding agents, Mr Kurtz. 'A remarkable person,' Marlow is told. 'Sends in as much ivory as all the others put together.' Kurtz is also a poet and an intellectual, the author of an eloquent report to the International Society for the Suppression of Savage Customs, on which he has scrawled: 'Exterminate the brutes!'

Marlow begins his journey, as Conrad had done, taking the long route around the rapids to 'Central Station' – the road that Stanley built from the port of Matadi to Stanley Pool. At Central Station, Marlow finds the talk is all about ivory. 'The word "ivory" rang in the air, was whispered, was sighed. You would think they were praying to it. A taint of imbecile rapacity blew through it all, like a whiff of some corpse. By Jove! I've never seen anything so unreal in my life. And outside, the silent wilderness surrounding this cleared speck of earth struck me as something great and invincible, like evil or truth, waiting patiently for the passing away of this fantastic invasion.'

At Central Station, Marlow learns that Kurtz is ill. He also hears rumours that he has descended into some kind of savagery. His journey to the Inner Station is delayed, but eventually he sets off upriver, just as Conrad did on his way to Stanley Falls.

Going up that river was like travelling back to the earliest beginnings of the world, when vegetation rioted on the earth and the big trees were kings. An empty stream, a great silence, an impenetrable forest. The air was warm, thick, heavy, sluggish. There was no joy in the brilliance of sunshine. The long stretches of the waterway ran on, deserted, into the gloom of overshadowed distances. On silvery sandbanks hippos and alligators sunned themselves side by side. The broadening waters flowed through a mob of wooded islands. You lost your way on that river as you would in a desert and butted all day long against shoals trying to find the channel till you thought yourself bewitched and cut off for ever from everything you had known.

The journey is filled with foreboding. 'Sometimes we came upon a station close by the bank, clinging to the skirts of the unknown, and the white men rushing out of a tumble-down hovel, with great gestures of joy and surprise and welcome, seemed very strange—had the appearance of being held there captive by a spell. The word ivory would ring in the air for a while—and on we went again into the silence . . . We penetrated deeper and deeper into the heart of darkness.'

Approaching Inner Station, Marlow, on the steamboat, observes Kurtz's house on the river through binoculars. On top of the fence-posts in front of the house, he glimpses what at first he thinks are ornamental knobs, but then discovers that each is 'black, dried, sunken with closed eyelids—a head that seemed to sleep at the top of that pole, and with the shrunken dry lips showing a narrow white line of the teeth'.

With a cargo of ivory and the ill Kurtz on board, Marlow returns downstream. Kurtz talks of grandiose plans, but dies on the way, whispering in despair, 'The horror! The horror!'

Conrad, too, set off downstream from Stanley Falls, piloting the steamboat *Roi des Belges*, with a cargo of ivory and a French agent for an ivory-collecting company, who died on board. A few years later, a Belgian officer in the Force Publique, who became station chief at Stanley Falls, gained notoriety for decorating the flowerbed in front of his house there with the heads of twenty-one women and children killed during a punitive military expedition.

Stanley himself returned to the Congo in 1887 at the head of an expedition to rescue Emin Pasha and his ivory hoard, under siege in the southern Sudan. He believed it would take him no more than two or three months after leaving the Congo River to traverse the unexplored Ituri rainforest of the northern Congo before he reached Emin. But his expedition encountered appalling hardship, and by the time he met Emin on the banks of Lake Albert in 1889, it was so depleted that he was obliged to leave the ivory behind.

In his account of the expedition, *In Darkest Africa*, Stanley railed against the depredations of the ivory trade in the Congo.

Every tusk, piece and scrap of ivory in the possession of an Arab trader has been steeped in human blood. Every pound weight has cost the life of a man, woman or child; for every five pounds a hut has been burned; for every two tusks a whole village has been destroyed; every twenty tusks have been obtained at a price of a district with all its people, villages and plantations. It is simply incredible that, because ivory is required for ornaments or billiard games, the rich heart of Africa should be laid waste at this late year of the nineteenth century, and that native populations, tribes and nations, should be utterly destroyed . . .

The elephant herds, too, were destroyed. But as ivory supplies began to dwindle, Leopold's fortunes were boosted by another commodity: wild rubber. The invention of the pneumatic tyre, fitted first to bicycles and then to motor cars in the 1890s, led to soaring demand for rubber. Just as the ivory 'rush' had produced a landscape of burned villages, refugee populations, slave labour and mass murder, so now the same happened with rubber.

To keep up the flow of rubber, company agents, backed by armed militias, imposed quotas on villagers. Villagers who failed to fulfil their quotas were flogged, imprisoned, and even mutilated, their hands cut off. Hundreds were killed for resisting the rubber regime.

The increase in rubber production was impressive. In 1890 the Congo exported 100 tons of rubber; in 1901, 6,000 tons. Leopold used this great wealth to fund a grandiose programme of public works, building palaces, pavilions and parks in Belgium, enjoying his reputation as a 'philanthropic' monarch. He also acquired a huge personal portfolio of properties in Brussels and on the French Riviera.

The public furore that eventually erupted over the Congo's 'rubber terror' forced Leopold in 1908 to give up his role as King-Sovereign of the Congo and hand over his empire to the Belgian state. But he remained fabulously wealthy, and when he died in December 1909 he possessed one of the largest fortunes in Europe.

In one final twist of fate, his death set off yet another ivory rush—the last of that era. It occurred in the Lado Enclave, a remote swathe of malarial swamp, forest and grassland, lying between the Congo and the southern Sudan that Leopold had fought tenaciously to acquire during the scramble for African territory by European powers in the

1880s. It had once formed part of Emin Pasha's domain. Then Britain had claimed it, along with all other territory bordering the upper Nile. Determined to gain a foothold on the Nile, Leopold had come to an arrangement with Britain to lease the Lado Enclave for the duration of his life and for six months after his death. Shortly after his death, as Belgian officials withdrew, leaving the area ungoverned until the British took over in June 1910, the hunters moved in.

The Lado Enclave abounded in elephants. A professional ivory hunter, 'Karamoja' Bell, travelled there in 1908, in the dry season when herds converged on the river: 'All the elephant for 100 miles inland were crowded into the swamps lining the Nile banks. Hunting was difficult only on account of the high grass. To surmount this one required either a dead elephant or a tripod to stand on. From an eminence others could generally be shot. And the best of it was the huge herds were making so much noise themselves that only a few of them could hear the report of the small-bore.'

Further inland from the Nile, he found a 'truly wonderful country', an area of high, rolling hills and running streams of clear water, with meadows of fresh green grass: 'In the far distance could be seen from some of the higher places a dark line. It was the edge of "Darkest Africa", the great primeval forest spreading for thousands of square miles. Out of that forest and elsewhere had come hundreds and hundreds of elephants to feed upon the young green stuff. They stood around that landscape as if made of wood and stuck there. Hunting there was too easy.'

Bell had dreamed of hunting elephant in Africa since boyhood in Scotland. He set himself a target of 1,000 elephants and pursued the idea relentlessly. To perfect his shooting, he studied the anatomy of an elephant as no hunter had ever done before. Determined to work out the best way for a brain shot, he sawed an elephant's skull in half and dissected the honeycomb cellular bone structure piece by piece to ascertain exactly where the brain lay. He did the same for a heart shot, crawling inside the carcass while his bearers drove spears in from the outside.

Bell witnessed the ivory rush begin:

All sorts of men came. Government employees threw up their jobs. Masons, contractors, marine engineers, army men, hotel-keepers and others came, attracted by the tales of fabulous quantities of ivory. More than one party was fired with the resolve to find Emin Pasha's buried store. It might almost have been a gold rush . . .

At first, they were for the most part law-abiding citizens, but soon this restraint was thrown off. Finding themselves in a country where even murder went unpunished, every man became a law unto himself, and the Belgians had gone. Some of the men went utterly bad and behaved atrociously to the natives, but the majority were too decent to do anything but hunt elephants. The natives became disturbed, suspicious, shy and treacherous. The game was shot at, missed, wounded or killed by all sorts of people who had not the rudiments of hunter-craft or rifle shooting.

Nevertheless, many fortunes were made before the British arrived. Bell himself shot 210 elephants in nine months, gaining five tons of ivory. He returned to the Sudan in 1912, assembled a steamboat he had brought from England and hunted along river islands. On his most profitable day, he collected 1,463 pounds of ivory from nine elephants he killed in seven-foot swamp grass near the Pibor River. That day's ivory earned him £900 at Hale's auction room in London.

# 13
## *Rivers of Ivory*

The flow of ivory from Africa in the nineteenth century reached around the world, to Europe, North America, India, China and Japan. African ivory was prized more than any other. It was finer-grained, richer in tone and larger than Indian ivory. East Africa on its own ranked as the world's largest source of ivory throughout the century. It produced what was known as 'soft' ivory that was white, opaque, smooth, gently curved and easily worked. West Africa tended to produce 'hard' ivory that was less intensely white, but glossy and more translucent.

In the new industrial era, the uses to which ivory could be put seemed unlimited. No other material responded so well to the cutting tools and polishing wheels of the Victorian age. It could be cut, sawed, carved, etched, ground or worked on a lathe. It could be stained or painted. It was so flexible that it could be turned into products such as riding whips, cut from the length of whole tusks. It could be sliced into paper-thin sheets so transparent that standard print could be read through it. An ivory sheet displayed at the Great Exhibition held at the Crystal Palace in Hyde Park, London, in 1851 was fourteen inches wide and fifty-two feet long.

Ivory in many ways was the plastic of the era. But it offered other highly valued qualities. It possessed a creamy, lustrous beauty that was unique. It was sensuously appealing to the touch. And its resemblance to the white-ness of skin was particularly alluring to the Victorian world which saw white skin as a symbol of status and purity.

Ivory workshops turned out a vast range of products: but-tons, bracelets, beads, napkin rings, knitting needles, door-knobs, snuff-boxes, fans, shaving-brush handles, picture frames, paper-cutters, hairpins and hatpins, and jewellery of all kinds. Ivory handles were fitted to canes and umbrellas, to hairbrushes and teapots. Ivory inlay work embellished mirrors, furnishings and furniture. A labour force of 600 was employed for ten hours a day in Aberdeen, Scotland, to man machines making ivory combs. Factories in Sheffield, England, imported hundreds of tons to make handles for cutlery.

Ivory featured in a wide range of musical instruments. It was used for the fingerboards of Spanish guitars; as pipe connections for Scottish bagpipes; as bridges for the violin and stops for the flute. Above all, it became the ideal material for piano keys.

The modern upright piano, invented by an Englishman, John Hawkins, in 1800, swiftly became the most popular instrument of the nineteenth century. 'Tickling the ivories' was both a favourite pastime and a social skill that was much admired. Ivory's smoothness, combined with its slightly porous texture, proved perfect for the pianist's touch. Factories in Germany, England and the United States turned out millions of piano keys to meet the demand. But it was the United States that took the lead. Production of pianos there rose from 9,000 in 1852, to 22,000 in 1860 to 350,000 in 1910. Ivory-working mills in Deep River and Ivoryton in the Connecticut River Valley dominated the trade in piano keys, supplying world-famous piano-makers such as Steinway. Each keyboard contained a pound and a half of ivory. In 1913, the United States used nearly 200 tons of ivory for making piano keys.

Ivory was favoured too by the makers of scientific instruments. Its whiteness, the ease with which it could be engraved, and its ability to absorb dyes and pigments, proved excellent for the production of instruments of measurement and dials. It was also durable, retaining its form and finish even with constant use. Navigational instruments, slide rules, telescopes and microscopes, all included ivory work.

The same qualities appealed to the makers of games such as dominoes, dice, draughts and backgammon. Ivory chess sets were also in vogue. But the game that eclipsed all others in its use of ivory was billiards. Billiards became fashionable among aristocrats in the eighteenth century; in the nineteenth century it became universally popular. At first, billiard balls were made from multicoloured hardwood. But in the early nineteenth century ball lathes in France and England were adapted for using ivory. No other material offered such a gratifying combination of touch and appearance, of density and elasticity, of internal balance and resistance to wear. Ivory balls impacting on one another produced an unmistakable 'click' which players relished. But the cost was high. Billiard balls and pool balls had to be cut from the dead centre of the tusk in order for them to roll properly; the tusk's black nerve canal was used as a centre line. The most that a sizeable tusk could produce was no more than four or five billiard balls. Workshops specialising in billiard-ball production, which sprang up in neighbourhoods in New York, London, Antwerp and Hamburg, consumed hundreds of tons.

Nothing, however, was wasted. Shavings, cuttings and scraps were bagged for further use. Ivory dust was boiled to make gelatine; it was burned to produce Indian ink and ivory black; it was used as fertiliser; as sizing for fabrics and paper; and as hair dye. Ivory shavings—sold at sixpence a pound in 1855—were boiled with water to produce a jelly; according to Victorian salesmen, it was 'the finest, purest, and most nutritious animal jelly that we know of', highly beneficial to invalids.

The ivory-carving workshops of Europe flourished as never before. Dieppe produced dynasties of master carvers famous for their statuettes and ship models. Napoleon and Josephine visited Dieppe in 1802, departing with ivory gifts, and French royalty continued to patronise the *ivoiriers* there in later years. Work shown by members of the Dieppe school won great acclaim at the Paris Exhibition in 1834.

Another renowned centre was established at Erbach in

Hesse. Carvers there developed a style known as 'Biedermeier', a name taken from a fictional character said to embody middle-class attributes of opulence and philistinism. Their speciality was ivory carved flowers, the flowers representing the language of love and ivory being seen as the most romantic of materials.

The traditional markets for African ivory, in the East, also multiplied their demand. India imported, as it had done for centuries, large quantities of East African ivory to make marriage bangles, which formed an integral part of Hindu and Muslim wedding ceremonies. India's elite far preferred 'soft' African ivory for luxury items to Indian ivory which was comparatively brittle and tended to discolour. As well as turning out a vast range of ivory objects for their own domestic markets, both India and China exported a multitude of figures, ornaments, chess sets, children's toys, puppets and chinoiseries to the West. Europeans were fascinated in particular by the artistic brilliance of Chinese carvers in producing 'devil's balls', concentric spheres containing a series of perforated ivory globes, moving freely, one inside another, in everdecreasing size. Japan's most distinctive ivory product was *netsuke*, small toggles which the Japanese used with string or cord to hold pouches and boxes suspended from their waistbands. Elaborately carved, *netsuke* became fashionable artefacts in the eighteenth century. In the nineteenth century huge quantities were exported to Europe and the United States.

The volume of ivory needed to keep pace with the world's demand was huge. In the sixty years from 1850 to 1910, Britain imported on average 500 tons of ivory each year. World consumption in the late nineteenth century reached about 1,000 tons. What this meant in elephant terms, according to contemporary estimates, was that 65,000 elephants were killed annually to satisfy the trade.

The scale of the slaughter eventually aroused international concern. It was increasingly evident that whole elephant populations were in danger of being wiped out. After spending three years travelling through central Africa, the German scientist and explorer Georg Schweinfurth warned: 'Since not only the males with their large and valuable tusks, but the females also with the young, are included in this wholesale and indiscriminate slaughter, it may be easily imagined how year by year the noble animal is being fast exterminated.'

# 14
## Jumbo and Friends

---

Whatever the plight of the African herds, in the world of entertainment elephants were accorded celebrity status. They were the stars of the circus ring and the zoo. Poets and writers told tales about them that delighted generations of children.

Hilaire Belloc set the tone:

> When people call this beast to mind,
> They marvel more and more
> At such a LITTLE tail behind
> So LARGE a trunk before.

The most famous elephant of the nineteenth century was an African bull named Jumbo. Born in southern Africa, Jumbo was taken to the Cape and sold to the Jardin des Plantes in Paris; and then, in 1865, at about the age of four, he was sent to the London Zoo in exchange for a rhinoceros. How he acquired the name of Jumbo remains uncertain. It was probably taken from the words mumbo-jumbo, a term used in the English language since the eighteenth century to denote a powerful African deity which various African tribes were said to worship. But, as a result of events that took place in London in 1882, Jumbo became the nickname for all future elephants and indeed the jargon for anything from aircraft to ice cream where size counted.

For fifteen years at his home at the London Zoo in

Regent's Park, Jumbo lived contentedly. Soon after his arrival, the zoo authorities purchased a second African elephant, a cow named Alice, which had been acquired by an Italian traveller in the Sudan. The two settled down amicably in adjacent stalls.

As he grew older, Jumbo became a star attraction. He was trained to carry up to six passengers at a time sitting on a wooden structure on his back, with a keeper leading him along the paths of the zoo's gardens. As well as rides, visitors enjoyed feeding him buns. For children, it was all a thrilling experience.

But in 1881, by which time he had grown to more than eleven feet tall and weighed six tons, his behaviour became increasingly unpredictable and aggressive. He attacked his quarters, driving holes through the walls and breaking off both tusks close to the jawbone. He was immediately withdrawn as a riding animal. But the zoo authorities feared that more drastic action might be needed.

In a letter to the Zoological Society's council, the zoo's superintendent, A.D. Bartlett, warned that staff members were at risk. Only Jumbo's keeper, Matthew Scott, seemed to be able to control him. 'I have no doubt whatever that the animal's condition has at times been such that he would kill anyone (except Scott) who would venture alone into his den.' As long as Scott could cope, the problem could be handled. However, he continued, 'in the event of illness or accident to the keeper (Scott) I fear I should have to ask permission to destroy the animal, as no other keeper would undertake the management of this fine but dangerous beast.'

While the council was deciding what action to take, an American showman, Phineas T. Barnum, offered to buy Jumbo and ship him to the United States. The council asked for £2,000 and Barnum readily agreed.

When news of the sale was published in *The Times* on 25 January 1882 there was uproar. Jumbo was regarded as a national institution. Newspapers denounced the deal. Questions were asked in parliament. Dissidents in the Zoological Society applied for a court injunction. Members of the royal family objected. Cartoonists depicted Jumbo with tears streaming down his face and Alice as the grieving wife. There were marches and protest songs.

The *Daily Telegraph* orchestrated a campaign to get

Barnum to cancel the purchase, offering £100,000 in return. The editor sent a telegram to Barnum appealing to him to reconsider.

All British children distressed at elephant's departure. Hundreds of correspondents beg us to inquire on what terms you will kindly return Jumbo. Answer prepaid, unlimited.

Barnum was unyielding:

Fifty millions of American citizens anxiously awaiting Jumbo's arrival. My forty years' invariable practice of exhibiting best that money could procure makes Jumbo's presence here imperative. Hundred thousand pounds would be no inducement to cancel purchase.

On the day that Jumbo was due to leave the zoo, the drama intensified when he refused to enter a large iron-barred crate intended to carry him to the London Docks six miles away and lay down in the road. A second attempt to get him there by leading him through the streets to the docks failed when he knelt down and would not move. The boat left without him. When he eventually sailed to America on 25 March 1882, accompanied by his keeper, Matthew Scott, a huge crowd went to see him off, offering him buns and even oysters and champagne.

Jumbo arrived in New York on 9 April. Cheering crowds lined the streets as a team of sixteen horses pulled Jumbo's cage from the harbour to Barnum's circus in Madison Square Gardens. For three years he entertained American audiences in the circus ring, leading in a procession of Asian elephants, over which he towered by three feet. Then, in September 1885, as he was leaving the circus grounds at St Thomas, a small town in Ontario, to return to his quarters, he was struck by a train and died within minutes. A post-mortem revealed his stomach to contain 'a hat-full' of English pennies, gold and silver coins, stones, a bunch of keys, lead seals from railway trucks, trinkets of metal and glass, screws, rivets, pieces of wire and a police whistle.

It was fictional elephants, however, as much as real ones, that captured the public's imagination. Rudyard Kipling's story of *The Elephant's Child* and how it acquired its trunk entranced generation after generation of children.

Kipling had travelled to South Africa in 1898 to escape the English winter. In Cape Town, he struck up a warm relationship with the mining magnate Cecil Rhodes, who arranged for him to travel northwards to visit the new British colony of Rhodesia which had been named after him. It was during that journey that Kipling reached 'the great grey-green, greasy Limpopo River' that later figured in *The Elephant's Child*.

Kipling returned to the Cape for the summer season every year from 1900 to 1908, staying in a beautiful Cape Dutch house in the grounds of the Groote Schuur estate beneath Table Mountain that Rhodes put at his disposal. It was there that Kipling wrote some of his *Just So Stories* for children. The fifth was *The Elephant's Child*.

Taking his cue from the recent discovery of primitive elephant fossil bones in Egypt, Kipling began his tale by explaining that in 'far-off times' the elephant had no trunk. 'He had a blackish, bulgy nose, as big as a boot, that he could wriggle about from side to side; but he couldn't pick up things with it.'

There was one elephant, however, an Elephant's Child, full of 'satiable curtiosity', forever asking questions, who changed all that. Wanting to know the answer to the question about what the Crocodile ate for dinner, the Elephant's Child received a spanking from his family, as he had done on so many previous occasions when asking questions. But a Kolokolo Bird, sitting in the middle of a wait-a-bit thorn-bush, offered him some advice: 'Go to the banks of the great grey-green, greasy Limpopo River, all set about with fever-trees, and find out.'

The next morning the Elephant's Child set out, eating melons along the way and throwing the rind about, because he could not pick it up. Sure enough, on the banks of the Limpopo, he came across a Crocodile. 'Will you please tell me what you have for dinner?' he asked him. 'Come hither,' replied the Crocodile, 'and I'll whisper.'

As the Elephant's Child put his head close to the Crocodile's 'musky, tusky mouth', the Crocodile seized him

On the charge, Boadicea, matriarch of the largest family in Manyara, named by Iain Douglas-Hamilton after the warrior queen. A few paces away she skidded to a halt and let out a mighty ear-splitting scream. The photographer, Lee Lyon, was subsequently killed while filming elephants in Rwanda. (Photograph taken in 1973.)

Carbon copy: baby elephants are born as almost perfect
miniatures of adult elephants ... They rarely stray more than
a few steps away from their mother, often seeking a secure
berth beneath her belly.

(*Above*) Infants form close relationships with brothers and sisters, displaying strong affection and spending hours in playful activity. (*Below*) Elephant families rest and sleep together, with the young carefully protected by adults within the herd, photographed by Joyce Poole.

Elephants spend much time on skin care, bathing regularly and showering themselves with mud and dust. Mud-wallowing is a favourite activity.

Two bulls in musth fight for supremacy, photographed
by Joyce Poole. Male contests for the right to mate occur
regularly and can end in injury or death.
Males also enjoy more friendly sparring.
Elephant greetings form a part of everyday life.

The trunk is a nose transformed into a limb of immense power
and mobility and sometimes needs a rest.

Elephants attach a particular significance to death. They show an acute interest in elephant bones, especially skulls and tusks, smelling them, turning them over, picking them up and carrying them off.

A group of Amboseli elephants on their morning journey to the swamps, with Mount Kilimanjaro in the distance, photographed by Cynthia Moss.

'Nature's great masterpeece,' wrote the English poet, John Donne,
'The onely harmlesse great thing.'

by his little nose and pulled and pulled and pulled, stretching his nose into a trunk. For three days, the Elephant's Child waited for his nose to shrink, but it never did. He discovered, however, that his new trunk could be used for a variety of purposes—for feeding, for scooping up mud, and for spanking other animals. Returning home, he demonstrated his new-found prowess by giving a thorough spanking to family members who previously had given him such a hard time. The other elephants were so impressed by what he could do with his trunk that they all hurried off to the Limpopo to get new noses from the Crocodile.

The most famous elephant of the twentieth century, Babar, also had African origins. Babar made his first appearance as an orphaned elephant adopted by a kind old lady in an illustrated book by the painter Jean de Brunhoff, published in Paris in 1931. He had started life the previous summer in tales told by de Brunhoff's wife, Cécile, to their children, simply intended to send them to sleep.

The de Brunhoffs had never been to Africa, but that summer they had listened to elephant stories brought back to Paris by an adventurous couple who had just returned from a long expedition there. One was de Brunhoff's cousin, Giselle Bunau-Varilla, a sculptress who had studied under Rodin; the other was Mario Rocco, a handsome Neapolitan aviator and former cavalry officer who had fled the threat of imprisonment by Mussolini for exile in Paris where he had met Giselle.

At Giselle's suggestion, they had embarked in 1929 on an elephant-hunting safari on foot in the eastern Congo. There they came across the wife of a Belgian administrator who had adopted a baby elephant. The Belgian couple were preparing to return to Brussels and rather than leave the elephant behind, they had decided to take it with them. On arrival in Brussels, the elephant went to live in the zoo where he was visited every day by the administrator's wife bearing croissants.

In the book *Histoire de Babar*, Babar flees the forest after hunters have killed his mother, heads for town and meets a rich old lady who takes care of him. He dresses in the latest fashion, acquires a suit of bottle green, a bowler hat and shoes with spats, and has his photograph taken. Having triumphed over adversity, he returns to the forest to be crowned king of the elephants.

Before he died from tuberculosis at the age of thirty-eight, Jean de Brunhoff wrote six more volumes about Babar, providing children with a whole manual on civilised life. His son Laurent carried on the tradition, producing thirty more volumes.

As for Mario Rocco and Giselle, they returned to Africa, built an art déco villa on the shores of Lake Naivasha in Kenya and established a family farm there. Their daughter, Oria, was the girl with long dark hair whom Iain Douglas-Hamilton met at a party in Nairobi and spirited away to his elephant camp at Manyara.

# 15
## *Safe Havens*

---

When Lieutenant Colonel James Stevenson-Hamilton, a
Scottish naturalist and hunter, arrived in the lowveld in
South Africa in 1902 to take up his appointment as warden
of the Transvaal's first designated game reserve, elephants
were believed to be virtually extinct there. 'Although in former days no doubt plentiful throughout the Lowveld,'
wrote Stevenson-Hamilton in his book, *Wild Life in South
Africa*, 'in 1902 the only indication of elephants in the present Kruger National Park were a few tracks in the neighbourhood of Olifants Gorge.' The number of elephants
thought to have survived was no more than ten. In 1905 a
group of about twenty elephants was found hiding in an
inaccessible gorge in the Lebombo hills.

The initiative to establish the game reserve had come
from Paul Kruger, the irascible, rough-mannered president
of the Transvaal during its days as an independent republic
before the Anglo-Boer War. A keen hunter in his youth, he
had amputated his own thumb after an accident with a
heavy four-pounder elephant gun, curing the resultant
gangrene by plunging his hand into the warm stomach of a
goat. At Kruger's instigation, the Transvaal parliament
resolved in 1895 to set aside land for a reserve in the eastern Transvaal, acknowledging 'that nearly all big game in
the Republic have been exterminated, and that those
animals still remaining are becoming less day by day, so

that there is a danger of them becoming extinct in the near future'.

The new park—Sabi Game Reserve—was formally designated in March 1898. But the Anglo-Boer War intervened and it was not until 1902 that its first warden, Stevenson-Hamilton, who had served as a cavalry officer in the war, took up his post.

The survival of the few remaining elephant herds in South Africa was still precarious, however. When the small herd in Addo in the eastern Cape, forty miles from Port Elizabeth, began raiding citrus crops on neighbouring farms, the authorities ordered their extermination. Extermination was seen as the only solution to the growing conflict between elephants and farmers expanding into territory where they had once roamed freely. A professional hunter was hired. One by one, some 120 Addo elephants were killed. By 1919 only sixteen were left. But this tiny band held out, hiding in impenetrable thorn thickets. After twelve years of intermittent incidents, the government gave in and in 1931 proclaimed Addo an Elephant National Park. Eleven elephants had survived the campaign.

Concern about the mass destruction of elephant and other African wildlife prompted Europe's colonial powers, now in control of most of Africa, to initiate a series of game protection measures. Largely on the initiative of Hermann von Wissmann, a former governor of German East Africa (Tanganyika), representatives of Britain, France, Germany, Portugal, the Congo Free State, Italy and Spain met in 1900 at the first International Conference for the Protection of Wild Animals and signed a convention 'for the preservation of wild animals, birds and fish in Africa'.

Colonial administrators imposed hunting restrictions,

banning hunting altogether in certain specified reserves. They outlawed a range of indigenous hunting methods practised for centuries, including the use of pitfalls, hamstringing, fire-rings and weighted spears dropped from trees. Where elephant hunting was permitted, it was subject to government licence.

In place of the white hunters who had wandered at will across Africa during the nineteenth century, a new breed of safari hunter emerged. One of the pioneers was Theodore Roosevelt, a former president of the United States, who set out on an African safari in 1909, shortly after his second term in office had ended, taking Frederick Selous with him as a guide. For the first part of the trip from Mombasa, he took the railway to Nairobi, sitting with Selous in the open on a seat built for him on the cowcatcher. 'It was literally like passing through a vast zoological garden,' he wrote in *African Game Trails*.

Roosevelt's safari took on the nature of a military expedition. Flying the Stars and Stripes, his Kenya party consisted of gun-bearers, tent-men, guards, syces and a column of 500 porters which stretched out for over a mile. From Kenya, he travelled to Uganda, moved northwards into the Lado Enclave, steamed down the Nile to Khartoum and arrived in Cairo nearly a year after setting out.

Along the way, Roosevelt and his son Kermit shot more than 500 animals, including eleven elephants. He killed his first elephant with only two shots. 'I felt proud indeed as I stood by the immense bulk of the slain monster and put my hand on the ivory. The tusks weighed a hundred and thirty pounds the pair.' Roosevelt celebrated the occasion by dining on elephant-trunk soup.

Though elephants were now protected from wholesale

slaughter, they faced a new formidable threat: economic development. Elephant herds were accustomed to roaming freely in search of food and water. For centuries they had used seasonal migration routes, well-defined elephant 'roads', beaten hard underfoot, and skilfully aligned to take advantage of the contours of hilly terrain. Some were reported to stretch for hundreds of miles. One elephant route in Uganda, running from the Murchison Falls area eastwards through Lira and on into Acholi country, was at one time regarded as the best road in the whole country.

But gradually the old routes and the old familiar territories were cut by the spread of modern agriculture, by the farms and villages of an expanding African population, by plantations and estates, and by roads and railways, some of which followed paths first engineered by elephants. Forests were cleared; waterholes and riverbanks were taken over by herds of livestock. Elephants raiding crops such as maize, sugar cane and bananas came into deadly conflict with farmers, as they had done in Addo. Farmers demanded protection from the authorities.

In Uganda, where nearly three-quarters of the land was considered elephant country, the British governor in 1924 ordered the establishment of a special unit to combat animal harassment of the local population. Organised on military lines, it was given the name Elephant Control Department. The name was changed in 1925 to Game Department, but its main purpose continued to be to control elephant depredations.

'There is still plenty of room in Uganda for both the [human] population and the elephants,' Captain Charles Pitman, head of the Game Department, recorded in 1925, 'but a comparatively large number of elephants must be destroyed each year to prevent them overrunning the country.' Each year an average of 1,000 elephants were killed during control work.

In neighbouring Tanganyika, the government at first offered professional hunters free licences to shoot twenty-five marauding elephants, but soon abandoned the scheme when unscrupulous hunters simply took the opportunity to select large tuskers whether they were causing trouble or not. Eventually, an Elephant Control Scheme was put in place, manned by white game rangers and African scouts.

This was still a time when elephant herds could be seen gathering in huge congregations. A professional hunter, George Rushby, recalled witnessing two similar occasions, one in the Mweru swamp in Northern Rhodesia (Zambia) in 1924, the other in the Ulanga Valley in Tanganyika in 1927. 'They covered an area some four or five miles long, and about one and a half miles in breadth. Within this area were scattered complete herds, varying from fifteen to thirty in each herd, and they all moved slowly forward in the same direction.'

Rushby estimated that there were about 700 elephants in each gathering. 'The herds were dotted about at roughly equal distances apart on both sides of the formation, and leading in the centre was an enormous single bull. A short distance behind him were three or four very big bulls and further back down the centre of the formation were other big bulls in twos and threes. Flanking these on both sides were the herds, and the young bulls in groups of threes and fours.'

A Uganda game warden, Rennie Bere, described seeing his first large gathering in 1934: 'Acres and acres of elephants—large elephants, small elephants, bulls, cows and calves; some were quiet, others playful, some feeding and others waving their trunks about in the air and screaming; and the whole time the interminable thunderous rumble of their bellies.'

But with the steady encroachment of human populations and modern agriculture, the main sanctuary for elephants became the reserves. Over a period of fifty years the reserves grew into a network of national parks scattered across Africa. The Sabi reserve was extended in 1926 to an area of 7,500 square miles and renamed Kruger National Park. Stevenson-Hamilton, who remained warden of Kruger until 1946, pioneering a whole new approach to wildlife conservation, estimated that by 1931 the elephant population there had increased to 135. In southern Tanganyika, an area of 21,000 square miles was set aside as a reserve and named after Frederick Selous who was killed there in 1917 in a skirmish between British and German forces near the Rufiji River. One by one, new parks were designated: Serengeti in Tanganyika in 1940; Tsavo in Kenya in 1950; Wankie in Rhodesia in 1950; Murchison Falls in Uganda in 1952. The number of parks and national reserves eventually grew to more than ninety, covering an area of 250,000 square miles.

Only one attempt was made during these years to domesticate African elephants. The initiative was taken originally by King Leopold of Belgium, while he was still in control of the Congo Free State. After learning that an orphaned elephant had been trained to carry out simple tasks by missionaries there, he appointed a Belgian army officer, Commandant Jules Laplume, to found an elephant-catching station.

Laplume set up his headquarters at Kira Vunga in the

rainforest of the Bas Uélé district where there were large numbers of forest elephants. His initial attempt to capture elephants, however, failed. He tried digging pitfalls, but succeeded only in injuring his captives. When he tried to separate calves from cows, the cows came charging back to the rescue. He eventually resorted to shooting cows, but found it difficult to keep orphaned calves alive. Nevertheless, by 1910 he had collected a corps of thirty-five young elephants. With trained elephants available, he devised new methods to rope calves without the need to shoot cows.

In 1925 the original station was transferred to Api and, in 1930, another station—the Station de Capture et de Dressage des Éléphants—was opened at Gangala-na-Bodio, on the banks of the Dungu River in Garamba National Park, close to the border with Sudan.

Visiting Gangala-na-Bodio in 1936, Armand Denis, a wildlife filmmaker, found the station being run with all the discipline and precision of a cavalry barracks. At sunrise every morning the Belgian flag was raised to a fanfare of trumpets while African *cornacs*—elephant-handlers— assembled for inspection. The *cornacs* then spent the day training their calves: tying their legs, forcing them to sit, teaching them to pick up objects with their trunks and preparing them to carry people on their backs.

Denis accompanied a group of elephant-catchers—*chasseurs*—setting out into the field with rifles and ropes to round up new recruits. Approaching a herd of several hundred elephants downwind, they identified a number of likely candidates and then, at a signal, rushed forward, firing in the air and shouting, deliberately stampeding the herd. For three hours they pursued them on foot, running alongside, waiting for a gap to open up, then slipping into the middle, and, with more shots and shouts, splitting the herd into smaller groups.

As the elephants tired, the *chasseurs* caught up with a young bull, slipped a rope around its hind leg, and pulled it towards a tree, tying it up and driving off adults coming to the rescue. Later in the day the young bull was tied between two trained elephants and led back to camp. It was dangerous work. Denis recorded that fifteen *chasseurs* had been killed in the previous thirty years.

Trained elephants were meanwhile used as draught animals for moving timber and for ploughing on surrounding farms, and for taking tourists for rides into the bush. Some were shipped abroad to zoos and circuses.

Despite such projects, knowledge of the world of African elephants remained limited. Much of it depended on the stories and anecdotes of game wardens and professional hunters. No scientific research had ever been carried out.

A new age of discovery was about to unfold, however, one that would not only open up the lives of living elephants, but also unlock the mysteries of the past.

# 16
## Ancestors

On an expedition to the Fayûm oasis in Egypt in 1879, the German explorer Georg Schweinfurth discovered a fossil site that was to provide the first clues about the ancient origins of the elephant. The oasis lay in a huge depression at the edge of the Nile Valley, an area of some 700 square miles which had sunk to a depth of 200 feet below sea level, exposing rock strata from the early Tertiary period 50 million years before. In the centre of the depression lay a lake known to ancient Egyptians as Lake Moeris. On an island towards the eastern end of the lake, Schweinfurth found specimens of fossil fish and of an extinct genus of whale. Returning to Fayûm in the winter of 1885–6, he collected more fossils from the escarpments on the north side of the lake.

Schweinfurth's work was taken up by British scientists. In a series of winter expeditions to Fayûm between 1901 and 1904, Charles Andrews from the British Museum discovered the remains of a large number of prehistoric animals. Among them were the bones of an animal that became known as Moeritherium, the Wild Beast of Moeris.

Small and pig-like, only two feet tall at the shoulder, with eyes and ears that were high up on its head, almost like a hippopotamus, Moeritherium bore little resemblance to modern elephants. It possessed no trunk, only a snout, rather like the Elephant's Child that Kipling described.

Yet, on careful study, Moeritherium turned out to be an early type of proboscidean, or trunked mammal, an ancient relative of the elephant which first appeared in the Eocene epoch about 50 million years ago. It possessed features that foreshadowed the evolution of the whole order of trunked and tusked animals that were soon to proliferate around the world. Its second incisor teeth in the upper jaw were much enlarged, showing that the formation of tusks had already begun; there were similar signs of tusks in the lower jaw.

British expeditions to Fayûm were followed by the Americans. In 1907 Henry Fairfield Osborn arrived in Cairo at the head of a lavish expedition financed by the American Museum of Natural History, bearing a personal letter of introduction from President Theodore Roosevelt to the Viceroy of Egypt, Lord Cromer. In the cool of winter, his party of fossil hunters set out for Fayûm, fifty miles south of Cairo.

Sixty camels strong [wrote Osborn], we wound our way past the pyramids of the western side of the Nile, skirted the fertile basin of the Fayûm, and struck southwards into the waterless desert until we reached the region that represented the ancient cradle of the elephant family. We at once set to work with a very superior force of Egyptian excavators from Kuft, under the direction of Mr Walter Granger and Mr George Olsen, two of the best fossil hunters in America, who stuck to their arduous post for nearly two months, until driven out by sandstorms and excessive heat. With their skilled aid, we soon discovered the burial sites of three of the early elephant dynasties.

Professor Osborn devoted the next twenty-five years of his life to compiling two huge volumes on the evolution of elephants, laying the foundations for modern understanding of their past. Entitled *Proboscidea*, the first volume was published in 1936 and the second in 1942. They contained a classification and description of no fewer than 352 different proboscideans, of which 350 were extinct. Later writers considered Osborn had carried his classification 'to rather an extreme', and reduced the number of valid species and subspecies by about half, to 164.

Moeritherium is the earliest known relative of the elephant. It was initially placed in the direct line of ancestry. But it turned out to belong only to a side branch, linked to a common ancestor further back in time. Moeritherium disappeared in the Oligocene epoch, 30 million years ago. But other proboscideans proved more successful. Growing to a huge size, with tusks, trunks and columnar legs, they spread out from their base in the swamps and fertile plains of northeast Africa, reaching every continent except Australasia and Antarctica. They developed a remarkable capacity to adapt to different climates and conditions, moving into tropical rainforests, mountain areas, dry plains and the inhospitable northern latitudes of Eurasia. When the Bering land bridge appeared, they crossed into North America; mastodon from the Mammutidae family migrating there formed their own American branch. For a period of more than 30 million years, this extraordinary collection of proboscideans reigned as the lords of creation.

The Elephantidae family, the youngest proboscideans produced by evolution, began to emerge in the Miocene epoch about 15 million years ago. The ancestor of this line, *Primelephas*, developed complex molars capable of grinding up rough vegetation; and it lost its small lower tusks, gaining larger tusks in the upper jaw. Among the descendants of *Primelephas* were the mammoths. Originating in Africa about 10 million years ago, they eventually spread out to other parts of the world. Woolly mammoths colonised northern areas of Europe and North America, growing a thick shaggy coat to protect them from the cold and snow. Closely related to the mammoths was the *Elephas*, a name taken from the Greek word for elephant first used by Herodotus in the fifth century BC. *Elephas* too originated in Africa and spread through the continent, but it thrived most readily in the southern part of Eurasia.

The most recent branch of the Elephantidae family was an African group which came to be known as *Loxodonta* and which began to emerge about 5 million years ago. The existing African elephant, *Loxodonta africana*, was a late arrival: it first appeared about 1½ million years ago.

One million years ago, some twenty species of proboscideans—mammoths, mastodons, gomphotheres, stegodonts, dinotheres and modern elephants—still inhabited most major land areas of the world. But climate change and natural disasters, together with human evolution, led to mass extinction. The African branch of *Elephas* disappeared about 35,000 years ago; the American mastodon died out about 8,000 years ago and the mammoths, hunted heavily by Stone Age peoples, became extinct about 4,000 years ago. Only two proboscideans survived: *Elephas* in Asia; and *Loxodonta* in Africa.

When drawing up the first scientific classification of animals in the 1750s, Carl Linnaeus, the Swedish botanist and zoologist, decided that the Asian and African elephant were members of the same genus and included both under the label of *Elephas maximus*. In 1797, however, Johann Blumenbach, the German anthropologist and physiologist, considered that the differences between them merited a separate classification and detached the African elephant from the genus *Elephas*. The following year the French zoologist Georges Cuvier named the genus for the African elephant *Loxodonta*, meaning 'slanting-toothed ones', on account of the lozenge shape of their grinding teeth. Included in the genus *Loxodonta* were a number of extinct elephants as well as the sole survivor, *Loxodonta africana*.

The differences are indeed considerable. The African elephant is generally heavier and taller: it weighs up to six tons (13,000 pounds) and reaches eleven feet or so at the shoulder. Its back is concave or saddle-shaped, whereas the Asian's back is convex or straight. The African has immense triangular ears that extend beyond the neck; the ears of the Asian are comparatively small. The African has a flat forehead, and carries its head high; the Asian has a twin-domed forehead, and carries its head low. The African's trunk is marked by repeated skin folds or 'rings' and at its tip are two finger-like projections; the Asian's trunk is smoother and ends in only one 'finger'. African elephants of both sexes usually carry tusks; with Asian elephants, tusks are confined mostly to males.

*Loxodonta africana* produced two living subspecies: *Loxodonta africana africana*, the savannah or bush elephant, and *Loxodonta africana cyclotis*, first classified in 1900, but commonly known since 1924 as the forest elephant. The forest elephant, now confined mainly to the rainforests of western and central Africa, is smaller than the bush elephant, usually less than eight feet at the shoulder. Its ears are shorter and more rounded, hence the scientific name *cyclotis*, meaning 'round ear'. Its tusks are straighter and thinner; and its ivory is harder and denser. Its small size gave rise to the myth of the existence of a 'pygmy' elephant in Africa, which endured for much of the twentieth century.

*Loxodonta*'s ancient origins have provided it with some unlikely living relatives. In the geological strata containing the bones of Moeritherium were the skeletal remains of an extinct version of sea cow. Upon further investigation, the order of the sea cow, *Sirenia*, or sirens, named because of the long association of sea cows with the mermaid legend, turned out to be related to proboscideans. Indeed, in the modern world, the last surviving sea cows—manatees and dugongs—are the elephant's closest living relatives. Aquatic mammals living in shallow coastal waters in the Atlantic and Indian Oceans, they share similar features in bone structure, in teeth and in general anatomy. As in the elephant, the molars of sea cows work their way forwards until they are shed and replaced by those behind. The male dugong carries tusks which, like the elephant's, are large incisors and grow continuously throughout its life. In female sea cows, as in female elephants, the vaginal opening is located on the lower belly, a feature typical in marine mammals.

Another unlikely relative is the hyrax, a small furry animal with a pointed muzzle, the size of a rabbit, that is common in much of Africa. Like the elephant, the hyrax walks on the soles of its feet. It shares similar features in the structure of both legs and feet; and in its upper jaw are two small tusks, formed from incisors.

Largely as a result of the work of Professor Osborn, more was known in the 1940s about fossil elephants than about living ones. On the horizon, however, was a new breed of scientist whose research would eventually transform understanding of the elephant world.

# 17
## Enter the Scientists

In his memoirs, Myles Turner, the warden of Serengeti National Park, recalls a conversation he had in 1959 with Colonel Rowland Jones, the retiring warden of Kruger National Park, while they were sitting one evening on the plains of Serengeti watching a panorama of buffalo, giraffe, wildebeest and impala unfold before them. Lowering his binoculars, the colonel remarked, 'Enjoy this while you can, Myles, because two things will ruin it in the end: tourists and scientists.'

Turner was puzzled. 'Tourists at that time were confined to a few wealthy Americans on hunting trips. And scientists in the field of wildlife were then unknown to me.' As Turner was to discover, all that was about to change.

A few pioneers had already begun work in Uganda. The first to arrive was John Perry, a Cambridge scientist invited by the Uganda Game Department in 1946 to undertake elephant research. Perry carried out the first modern dissection of an African elephant—nearly 300 years after an Asian elephant had been dissected in Dublin—and he continued with a ground-breaking study on reproduction. He noted in 1952, after completing his work, what little progress had been made with scientific research. 'Although the elephant has always excited interest, surprisingly little was known about it beyond the scattered and often fanciful descriptions of its habitats published by hunters. There is very scant knowledge of the details of the anatomy of the African elephant and still less of its physiology.'

Perry was followed in 1956 by a group of American ecologists whose initial project was to study the problem of tree damage caused by elephants in the Murchison Falls National Park. Among them was Irven Buss, the first scientist to counter the notion held by generations of hunters and wardens that elephants were led by 'herd bulls' or 'sire bulls'. Buss suggested in 1961 that elephants assembled in family units of closely related cows and their offspring.

Further research work on elephants was carried out by the Nuffield Unit of Tropical Ecology, established in Uganda in 1961 by Richard Laws, previously an authority on whales. The volume of research material steadily accumulated, providing insights into the numbers, distribution and movements of elephants; their growth, weight, height and age; their diet and habitat; and their population dynamics based on new figures on reproduction and mortality rates.

In Tanzania, meanwhile, another research project was established in 1961 by John Owen, the director of national parks. Known initially as the Serengeti Research Project, it had modest beginnings: two scientists began work, one on wildebeest, the other on zebra. In 1964 its name was changed to the Serengeti Research Institute and an influx of eager young scientists arrived.

'Word must have got back to the seats of learning about the opportunities in the Serengeti, because from then on we were inundated with scientists of many nationalities,' wrote Myles Turner, the warden. 'In those days there was little question of research being geared to Park management, and a determined smash-and-grab raid for PhDs was started by youngsters who regarded the Serengeti and its animals as a vast natural laboratory to be looted at will.'

A cluster of new buildings sprang up four miles from the park headquarters at Seronera to accommodate the work of biologists, ecologists, ethnologists, foresters and mapmakers.

'The arrogance of some of these scientists—with the ink hardly dry on their graduation papers—was unbelievable,' wrote Turner. 'I once heard them described at a research meeting, chaired by a very eminent visiting Oxford professor, as "these brilliant young men at the height of their creative powers"! They obviously believed in this assessment.' Turner found it difficult to accustom himself to this new breed of scientist, deplored 'their eccentric lifestyle, speeding around the park with long hair and odd clothes' and complained of their overweening confidence 'in their ability to find the answer to everything about wild animals'.

Iain Douglas-Hamilton first arrived in Seronera in its early days in 1963 as an undergraduate student from Oxford on a summer vacation; his task was to assist a scientist working on wildebeest. After taking a degree in zoology, he had hoped to return to Serengeti to study lions, but was told that a renowned American zoologist, George Schaller, was already working on lions. Instead, he was offered some unpaid research work on elephants in Manyara, a national park to the east of Serengeti.

Douglas-Hamilton's subsequent studies in Manyara produced major breakthroughs in the understanding of elephant behaviour. He established the main features of their social organisation, confirming Irven Buss's earlier proposition about the importance of family units. And the research techniques he pioneered to identify individual elephants proved invaluable to other elephant biologists.

Douglas-Hamilton's work in Manyara was taken up by Cynthia Moss in Amboseli, a small national park in southern Kenya. A former journalist on *Newsweek* magazine based in New York, Moss began working on elephants in 1968 as an assistant to Douglas-Hamilton. In 1972 she established her own research project in Amboseli, setting up a tented camp in a palm grove at the base of Mount Kilimanjaro, with breathtaking views of its snow-capped peak twenty-five miles away. Over years of observation, Moss came to know virtually the whole elephant population in Amboseli, a relatively self-contained group then numbering some 600 elephants living in an area of 150 square miles. On a scale never before attempted, she recorded births and deaths, calving intervals, family connections and social dramas, providing intimate insights into elephant family life. Her books and films made Amboseli's elephants the most famous in the world.

Moss was joined in Amboseli by Joyce Poole, a young American biologist, who concentrated on male society there, again with ground-breaking results. In 1985 Poole and another American biologist, Katherine Payne, who had spent fifteen years studying the songs of humpback whales, embarked on a new project in Amboseli which opened the way into the hitherto secret world of elephant communication.

Work on the study of elephant movements also advanced in leaps and bounds. The development of modern drugs made it possible to immobilise adult elephants for marking operations and then to revive them unharmed as a matter of routine. At first, elephants were marked with painted numbers and given ear tags, but the paint soon rubbed off or was obscured by mud and dust, and the ear tags were easily damaged. They were then fitted with colour-coded and notched collars made of machine belting, but collars often proved difficult to see; and elephants sometimes disappeared from view for months at a time.

With technological progress, biologists were able to fit miniaturised radio transmitters to collars, enabling them to track elephants from the ground or from the air for as long as batteries could last. In Douglas-Hamilton's first experiment on the ground in Manyara, he managed to keep track of an elephant for twenty days. In later years, it became possible to track elephants for periods as long as three years. In Rhodesia (Zimbabwe) in the late 1970s, Rowan Martin followed several dozen elephants from a single population for several years. When analysing the results, he noticed that certain pairs of families, while separated by miles of forest, were still able to coordinate their movements for days and sometimes weeks at a time without meeting, making the same changes of direction.

The ultimate development in tracking elephant movements has come with the use of satellite-based global positioning systems. In a project launched by Douglas-Hamilton in Kenya in the 1990s, elephants fitted with transmitters have been sending a flow of information via satellite, giving a fix on their position day and night, which is then stored on memory chips on their radio collars. All that is required to retrieve it is for an aircraft with receivers to fly over the study area once every few months.

The use of DNA technology has brought further advances. Not only have DNA samples provided scientists with evidence about paternity and relatedness among elephants, they also offer clues about the origin of tusks, a potentially vital factor in the campaign against illegal ivory dealing.

Alongside modern technology, field biologists have still persevered with more basic methods. During the 1980s Richard Barnes, a Cambridge scientist, set out with his wife, Karen, to ascertain elephant numbers in the rainforests of Gabon by counting their dung. No one had any idea of the real number of 'forest' elephants. They were impossible to count from the air and ground counts were difficult and unreliable. Elephant dung, however, was easier to find than elephants. And Barnes believed that he could provide a worthwhile assessment.

For two years, he cut his way through the Ogooué-Ivindo forest of northeastern Gabon, marking out twelve-mile-long transects, wading through rivers and marshes that lay in his way, recording the details of elephant dung he found. Each pile of dung was dutifully recorded, its position noted, its age estimated. Age was divided into four categories ranging from 'A' for dung that was 'intact, very fresh, moist, with odour', to 'D' for boluses that had 'completely disintegrated'. The results were impressive enough for other teams of 'dung-runners' to be established in Congo-Brazzaville, Congo-Kinshasa, Cameroon and the Central African Republic.

Further studies of forest elephants in the Central African Republic were made during the 1990s. Andrea Turkalo, a former biology teacher from New York, spent a decade working from a rough-hewn platform built at the edge of an opening in the rainforest at Dzanga Bai, observing forest elephants as they emerged briefly to drink and dig for minerals. In 2000 she embarked with Katy Payne on an

Elephant Listening Project at Dzanga Bai, intended partly to make further progress on decoding elephant language, but also to determine whether, by eavesdropping on their communications with networks of microphones, it was possible to estimate the number of elephants living in the rainforest and to gauge their behaviour.

Whether by using satellite technology or by sifting through dung, year by year biologists succeeded in extending the frontiers of understanding of elephants, uncovering one of the most complex societies of the animal world.

# 18
## *Family Life*

Baby elephants are born as almost perfect miniatures of adult elephants. After a gestation period of twenty-two months, they arrive in a foetal sac weighing on average around 120 kilos (260 pounds), with a short trunk, ears that resemble the map of Africa, pristine toenails and light patches of red or black body hair.

Within half an hour they struggle unsteadily to their feet, often falling over, patiently helped by their mother prodding gently with forefeet and trunk. Their first instinct is to seek out the teats on their mother's breasts which lie, remarkably similar in size and shape to human breasts, between her front legs. Suckling with their mouths, they consume up to ten litres (seventeen pints) of milk a day, depending on milk alone for the next six months.

Within hours of birth they are able to walk well enough to keep up with the family herd on its continuous amble in search of food. But otherwise they remain largely helpless, needing the constant care and attention of their mother. Rarely do they stray for more than a few steps away, spending long periods in physical contact, leaning or rubbing against her, touching her with their trunks and enjoying frequent spells of suckling.

Cows display great affection for their calves, constantly caressing them with their trunks to give reassurance, murmuring softly and providing them with a secure berth beneath their belly. The deep attachment they feel is shown most poignantly at times of death.

In Amboseli, Joyce Poole came across a calf that had been born dead or had died shortly after birth. The mother, a cow named Tonie, repeatedly nudged it with her feet and finally rolled it over several times. While the rest of the family moved on, Tonie refused to abandon her calf, standing vigil in the hot sun on a barren plain, fending off vultures and other scavengers for two days and nights, until lions dragged the corpse away.

'As I watched Tonie's vigil over her dead newborn,' wrote Poole, 'I got my first very strong feeling that elephants grieve. I will never forget the expression on her face, her eyes, her mouth, the way she carried her ears, her head and her body. Every part of her spelled grief.'

In Uganda, Rennie Bere, a game warden, described finding a cow carrying a dead calf which he judged by its smell to have been dead for three or four days. 'She placed this gruesome little object on the ground beside her whenever she wanted to feed and drink and did this several times while I was watching; although this made her travel slowly, the rest of the herd invariably waited for her.'

Clive Spinage, a British biologist, recorded an incident he witnessed when accompanying a game warden to shoot an injured calf, about eighteen months old. One of its legs had been caught in a snare, turning the foot into a swollen, festering mass. The calf was managing to hobble along behind its mother, while the rest of the family fed slowly ahead, but it had no hope of surviving. At the first shot, the calf was bowled over, mortally wounded, but did not die immediately.

The dam, screaming with alarm, instantly rushed to the rescue. Trumpeting with rage she frantically pushed the calf to its feet and tried to lead it away. Lolling against her legs, it staggered a few paces and collapsed again. Tusks streaked with blood, the mother desperately tried to raise it until, helpless, she stood over it as it lay there. Then, just after it died, she raised her trunk high into the air and gave vent to a penetrating wail of anguish. And it was a wail, not a trumpet, which echoed forlornly over the countryside.

As well as receiving help and affection from their mothers, calves are tended by an array of aunts, sisters and cousins who take a maternal interest from the moment of their birth, often crowding round the event in excitement. Young females in the family delight in their role as baby-sitters, standing over calves while they are sleeping, retrieving them if they wander too far and rushing to their aid at the slightest cry for help. When calves get stuck in mud, fall into holes or trip over logs, their loud squeals bring baby-sitters running from all directions.

Katy Payne witnessed a female calf fall into the deep end of a man-made watering trough, with a wild bellowing scream. Instantly, an aunt and two siblings ran to the calf's aid. Falling on their knees beside the calf, they reached their trunks under her belly to try to lift her out.

'As they struggled, their screams, bellows and rumbles were added to hers. Instantly more help came out of the forest . . . Thirteen mature female elephants ran forward and drew the infant with their trunks to the shallow end of the trough. Safe and pampered, she clambered out amid a pandemonium of reassuring rumbles.'

In his early experiments with immobilising elephants to

fit them with radio collars, Iain Douglas-Hamilton selected an eight-year-old male calf, assuming he was past the age to provoke a strong response from his surrounding family. He belonged to a well-integrated family unit consisting of five mature females and their calves, numbering sixteen in all. Several other families were close by at the time.

When the calf slumped to the ground, the mother rushed forward, extending her trunk and flapping her ears forward. All the other cows in the family converged on the calf, their ears and heads down, trumpeting and growling. Then three other families charged in, forming a phalanx of sixty-seven elephants in all. 'The young milled around, backed into each other and charged into space in a vacuum release of aggressive drive, all the while trumpeting, groaning, bellowing and growling in an indescribable and continuous confusion of noises, some of them remarkably human.'

After repeated attempts had been made to raise him, the calf eventually stayed on its feet. 'All the females of the five family units mixed once more in a huge conglomeration curious to find out how he had fared. Many came up and touched him with their trunks.'

In Botswana two filmmakers, Derek and Beverly Joubert, witnessed the rescue of a calf stuck in a mud hole which had been abandoned by its own family fleeing in fear and disarray. Soon afterwards, another group of elephants arrived and tried to release the calf from the mud. They too took off abruptly, leaving the calf screaming piteously. But then they returned and dug and pulled the calf out of the mud. It was led away, stumbling, under the belly of a matriarch who subsequently adopted it as her own.

Calves enjoy a long childhood. Like humans, they need

an extended period of nurturing to learn about the world around them and to absorb the knowledge of family elders. At birth, the weight of their brain is only 35 percent of the weight of an adult brain—a feature similar to the human brain which weighs at birth only 26 percent of its adult weight. In most other mammals the brain weight at birth is 90 percent of adult weight.

Baby elephants do not even know at first what to do with their trunks. They fiddle with them, swing them back and forth and whirl them around in a circle. Sometimes they suck them, like human babies sucking their thumbs. They bite at grass, pulling out clumps with their mouths, before discovering that the trunk can be used to pick up food items and reach water on the ground. It takes a year for them to be able to drink with their trunks with any accomplishment.

The process of weaning is gradual. Calves remain highly

dependent on their mother's milk for two years. They continue suckling for another two years or so, until the arrival of a new calf, and often longer in some cases. They sometimes experiment with food by taking it from their mother's mouth, gradually learning what plants are palatable.

They become accustomed to the endless wandering of the herd in search of food, finding their way over rough and difficult terrain, up hillsides and into swamps. They learn to swim at an early age, crossing rivers on the upstream side of their mothers, alternately submerging, then bobbing back up.

Like Kipling's Elephant's Child, they are often insatiably curious, venturing out to explore new things. George Rushby, a game warden in Tanganyika, once watched a herd move slowly in his direction upwind as he hid on a low mound behind a small bush.

The nearest one was a cow, who passed within nine or ten feet of me, and with her was a very young calf. The cow stopped a few paces beyond me, and the calf, which was between the cow and myself, walked up the mound towards me and with its very small trunk felt over my chest and face for a few seconds. It then turned and ambled to its mother and blew into her mouth with its trunk, as if trying to convey to her the scent of this strange thing it had found. The cow gave the calf a light cuff with her trunk and started to move on.

They form close relationships with older brothers and sisters, spending hours in playful activity. They race about at random, with their ears out, their tails curled over their backs, making mock charges at all and sundry, even birds and butterflies. They delight in climbing on each other, particularly in mud wallows, often ending up in a heap of wriggling bodies. Other favourite pastimes include pushing matches, trunk wrestling and head-to-head sparring.

Family life is highly organised. Each family is led by its oldest female, the matriarch, who takes decisions about where to go, when to move and when to sleep. Matriarchs serve as the repository of the family's knowledge and wisdom, knowing from long experience where water for drinking is to be found, when trees are fruiting, where grass is lush, what to do in drought conditions and which areas are dangerous because of human encroachment. Other members of the family—related adult females and their off-spring ranging from newborn calves to adolescent males and females up to about ten years old—follow her lead.

Daily activities are coordinated. Family members in groups of ten or fifteen feed, walk and sleep together. When resting, they gather in a tight-knit group, touching and leaning on one another. They particularly enjoy wallowing in mud together, lying down on one side, then the other, with mud over their eyes and ears.

Adults as well as calves engage in playful activity, spraying water about with their trunks and tossing vegetation in the air. At moments of abandon, they run about in a loose and floppy manner with their heads down, their ears flapping and their trunks waving about. Cynthia Moss once witnessed a congregation of 200 elephants indulging in 'floppy running', trumpeting loudly as they raced across an open pan.

In the event of trouble, elephants act readily in mutual help out of loyalty and compassion. When alarmed, they quickly bunch together, with the matriarch taking a prominent position while calves are protected within the phalanx. When under threat, cows may charge on their own, but if the danger warrants it, the whole family sallies forth, bellowing. Numerous cases have been reported of cows propping up or supporting wounded companions, trying to get them away from danger, showing great reluctance to leave them behind.

While young males leave the family at puberty at about the age of twelve, females remain in the family all their lives, forming part of an intricate hierarchy based mainly on age. They begin to breed at about the age of eleven, producing their first calf at about thirteen years old. They form lifelong friendships, spending much time caressing one another with their trunks, sometimes standing face to face with trunks entwined.

Beyond their immediate circle, family groups maintain strong ties with other elephant families. Families within this wider bond or kinship group sometimes feed and travel together; their calves play together. Larger groupings also occur. During the course of a lifetime, a matriarch has a passing acquaintance with scores of individual elephants.

Family reunions invariably inspire displays of affection. Even after short periods of separation, while feeding apart, family members greet one another with loud rumbles, head postures and ears spread out. Families meeting after days apart make a special occasion of it.

In her book *Elephant Memories*, Cynthia Moss describes the reunion of two matriarchs from the same bond group, running towards each other after five days apart, rumbling, screaming and trumpeting. 'Both elephants raised their heads up into the air and clicked their tusks together, wound their trunks around each other's while rumbling loudly and holding and flapping their ears in the greetings posture. They whirled around and leaned and rubbed on one another. Meanwhile all the other members were greeting each other with much spinning, backing, urinating, ear-flapping, entwining of trunks, and clicking of tusks.'

Moss wrote of how after years of watching elephants, she still felt a tremendous thrill at witnessing a greeting ceremony. 'Somehow it epitomises what makes elephants so special and interesting. I have no doubt even in my most scientifically rigorous moments that the elephants are experiencing joy when they find each other again. It may not be similar to human joy or even comparable, but it is elephantine joy and it plays a very important part in their whole social system.'

# 19
## *Mating Pandemonium*

When young bull elephants leave the comforts of the family herd at about the age of twelve, they enter a separate world of solitary lives. Some remain in the vicinity of their old families, tagging along in the distance; others move off to hang around the fringes of other families. As they grow older, they gravitate towards all-male groups but never stay for long. They indulge in frequent trials of strength, mock fights that serve to establish a hierarchy long remembered in adulthood. Though teenage males are sexually mature, they stand no chance of mating. Only in their late twenties do they begin to compete with older bulls for females and even then the most they can hope to achieve is a 'sneak' copulation.

The turning point for bull elephants comes at about the age of thirty when they begin to exhibit periods of heightened sexual aggression. The condition is known as 'musth', a name taken from the Urdu word meaning 'intoxicated'. Centuries ago, musth was identified as a common feature in the sexual cycle of Asian elephants, but it was thought not to occur in African elephants. Even in the 1960s, pioneers in the reproductive biology of African elephants such as Irven Buss and Richard Laws maintained it did not exist.

In 1976, however, Joyce Poole, then a nineteen-year-old undergraduate working in Amboseli with Cynthia Moss on a study of bull elephants, noticed on separate occasions a number of individual bulls whose penis sheaths had turned a greenish colour and who were constantly dribbling urine. Their faces were marked by the dark stains of secretions oozing from swollen temporal glands. Poole and Moss speculated at first that the affliction was a symptom of disease and they named it Green Penis Disease. In subsequent observations they noted that elephants with Green Penis Disease were particularly aggressive. But it was not until January 1978, when Poole was shown an article about musth in Asian elephants, that she realised what Green Penis Disease was.

With considerable fortitude, Poole went on to make the first comprehensive study of males in musth. The work was known to be dangerous. Charles Darwin in his *Origin of Species*, published in 1859, had warned that 'no animal in the world is as dangerous as an elephant in musth'. In the early stages of her investigation, Poole often returned to camp white-faced and shaken by her encounters with aggressive bulls. 'I was still a relative novice at discerning the moods of elephants, and on many occasions my own miscalculations brought me dangerously close to death,' she wrote in her book *Coming of Age with Elephants*.

In order to measure testosterone levels, she needed to drive up to musth bulls in her vehicle quickly enough to collect samples of urine deposits before they soaked into the ground. The difficulty was that the only time musth bulls dribbled urine in any useful quantity was when they were annoyed. One notoriously aggressive bull set out to terrorise her, coming for her vehicle from more than 400 yards away, and stalking her in revenge even when she retreated a long distance away. But she was eventually rewarded with impressive results.

The sexual cycle of male elephants is a yearly one. Generally, large males over thirty spend three months or so in musth and nine months in 'retirement' leading peaceful and solitary lives in 'bull areas' away from cows and calves. Their musth periods are spaced out throughout the year so that at any point some musth bulls are active in the field. When in musth, they constantly seek out females in oestrus, moving from one family herd to another, fending off rivals and engaging in fights, to the death if necessary, to attain matings. Their aim is to mate with as many females as possible.

They carry themselves differently, their heads held high, striding out purposefully, waving their ears, all the while dribbling urine and secreting a dark viscous fluid from their temporal glands. They exude a pungent, sharp smell, detectable from afar and send out low-frequency but loud musth rumbles which can travel over long distances. Their testosterone levels are at least four times higher than in non-musth periods.

In the presence of musth bulls, younger males move off. Bulls not in musth, while still capable of impregnating cows, also give way even though they may be larger and older. Lower-ranking bulls in musth are likely either to take off in the opposite direction or to stop exhibiting their musth condition altogether, lowering their heads. Considerable efforts are made to avoid fighting over females. Only musth bulls of similar rank engage in combat.

Females of all ages, however, show great excitement. Coming across the scent trail left by a musth bull, they stop to smell, sometimes rumbling and urinating. Hearing a musth bull rumble, they answer with a female chorus of rumbles. They can come into oestrus at any time of the year for periods that on average last about four days. Their aim during those four days is to mate with the best male available and they clearly prefer musth bulls. But they are surrounded by many other candidates and spend much of their time escaping from ardent young males, usually outrunning them. Once they have conceived, they will not come into oestrus again for three or four years. Male opportunities to mate are accordingly limited.

Musth bulls approach a family of females in a nonchalant manner, holding their heads slightly lower with their trunks often draped over a tusk. Testing for signs of oestrus, they move from one female to another, probing between their hind legs, taking urine samples with the tips of their trunks which they then place in their mouths. If they find nothing of interest, they soon head off for another family.

The mating routine, when it occurs, is swift. A cow in

oestrus walks away, but with her head turned slightly, looking back over her shoulder. The bull sets off after her, releasing his penis from its sheath. She begins to run, but without much determination. He quickly catches up, reaching out with his trunk. As he places it along her back, she stops.

Using his head as a lever, he rears up, placing his front feet just behind her shoulders, but leaning back so that his weight—as much as three times more than a cow—falls on his own hind legs. His penis, extended to its full length of three to four feet and weighing some 25 kilos (sixty pounds), curves into a S-shape, with the last foot or so moving up and down and from side to side almost independently. With this mobile penis, he searches for the cow's vulva lying low between her legs, its opening facing the ground. With the tip of his penis inside her vagina, he thrusts upwards, remaining in position for less than a minute, before dismounting.

General excitement now follows. The cow lifts her head and lets out a deep, pulsating rumble. Members of her family rush forward, rumbling, bellowing and touching the couple with their trunks. It becomes a family event—what Cynthia Moss calls 'mating pandemonium'.

The bull remains 'in consort' for a day or two, guarding the cow against the attention of other males in the vicinity. If he becomes distracted, 'sneak' copulators take their chances. But all lose interest once the oestrus period is over. Musth bulls soon move on in search of other likely females.

When fights occur, they sometimes last for hours. Before engaging, musth bulls pace up and down, manoeuvre and sidestep, ensuring that they keep facing each other. To demonstrate their aggression, they tusk the ground, uproot bushes and throw logs in the air. Then they lunge at each other, colliding with a tremendous thud and clank of ivory, each attempting to twist the other off his feet. Once down, an elephant is highly vulnerable. When one clash is over, they resume the same manoeuvres before engaging again. The contest goes on and on until one of them runs away. Cynthia Moss in Amboseli once recorded a fight lasting for more than ten hours.

The results are sometimes fatal. In Tsavo, park rangers observed a fight develop between two bulls, apparently to determine dominance and 'possession' of a cow in oestrus. They started by sparring but were soon locked in a full-scale battle. One bull lunged viciously, driving one of his tusks through the roof of his opponent's mouth and the other into his chest. The force of the blow lifted his opponent's forelegs off the ground. The stricken bull tried desperately to disengage but in doing so exposed his flank. Taking swift advantage, the attacking bull plunged both tusks into the flank just behind the shoulder, killing his opponent in seconds. Bleeding profusely from gashes on his trunk and forelegs, the victor moved off briefly to drink from a nearby pool, but soon returned in a rage, trumpeting loudly, and charged at the dead elephant, ramming his tusks full length into its head. Then he stood guard over the corpse for six hours.

On a hunting expedition in southern Tanzania, Mike Carroll, a friend of Irven Buss, witnessed a fight between two bulls of similar ferocity. Starting as a scuffle, it developed into a deadly duel. Again and again, the two bulls charged at each other, inflicting deep wounds. With their tusks engaged, one bull gave a mighty twist with his head, managing to break off one of his opponent's tusks. 'One

Tusk' struggled on, bleeding heavily from wounds on his chest, but the fight was now uneven.

Suddenly he dropped his head, turning it at the same time, thus aiming his tusk at his opponent's throat. Dropping his head allowed his opponent's tusks to go high on his head, tearing a terrible gash above his eye and ripping a great hole in his ear. But that one tusk thrust home and caught his opponent in the throat, going deep. With a tremendous heave One Tusk raised his head, lifted the impaled bull off his front feet, and ripped a great hole in his neck. As he caught his balance his guard dropped, and again the one tusk went home, this time through the trunk and deep into the head. Both elephants went down on their knees.

The one tusker immediately arose, tossed his head, and again thrust deep into the head of his opponent. With this blow the stricken bull went over on his side, feet flaying in the air. One Tusk quickly stepped around and repeatedly drove his tusk into the fallen bull's back, all the while trumpeting and screaming.

Bleeding badly from chest wounds, One Tusk continued to gore the dead bull for nearly an hour before moving off to a stream to drink. Climbing up a steep incline from the stream, he fell dead.

# 20

## *Body and Mind*

---

Though cumbersome in appearance and by far the largest land animal on earth, African elephants are none the less remarkably agile. They can move faster than any human. They can climb steep hillsides and negotiate their way along narrow ledges and over rocky outcrops. They are masters of the art of concealment and can tread in total silence. They can swim with ease across rivers and lakes, though often they prefer to walk along the bottom with their bodies submerged and their trunks held above the surface like periscopes.

Their range of habitats is equally diverse. They thrive as much at sea level as on the mist-enshrouded slopes of mountains such as Mount Kenya, 12,000 feet above sea level. They have adapted as readily to living in the deserts of Namibia and Mali as in the dense rainforests of central Africa or in winter conditions on the South African highveld. They have even ventured deep underground inside Mount Elgon, a dormant volcano straddling the Kenya–Uganda border, feeling their way in pitch darkness along cavern walls with their trunks to dig for mineral salts.

They possess many distinctive features. Their ears are the largest in the world. They have the biggest brains of all land mammals. Their molar teeth are the size of house bricks. More surprisingly, their feet are constructed in such a way

that they walk on tiptoe. But
what is unique about them is
their trunk. No other animal has
a trunk.

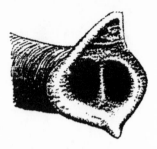

The trunk is a nose transformed into a limb of immense
power and mobility. Consisting
entirely of a mass of interlacing
muscles, it can bend, fold or curve in any direction. It developed during the course of evolution as a fusion between the
nose and the upper lip to enable elephants, with such large
bodies, long legs and short necks, to pick up food and water.
Early naturalists described it as 'the elephant's hand'. It is
used as a hose to spray water; as a soundbox for producing
trumpets, screams and squeals; and as a weapon of attack
capable of hurling humans thirty yards.

It is also remarkably sensitive. So precise is the movement
of the two 'fingers' at the tips of the trunk that elephants
can pluck a leaf or a small fruit from a cluster of foliage; or
pick up a single seed from the ground; or remove a piece of
grit from an eye; or scratch an ear.

With similar finesse, elephants use their trunks to com-

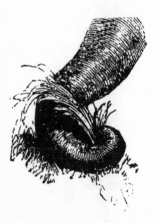

fort, to reassure and to show
affection. They spend much
time caressing calves and touching other family members in
an intimate manner, sometimes
standing face to face with trunks
entwined. They greet each other
by placing the tips of their
trunks into each other's mouths.
And they communicate inten-

tions with their trunks, holding them in different positions.

The trunk is also used as an early warning system. At the slightest sign of danger, elephants raise their trunks aloft, testing the air for scent. Their sense of smell is acute. With relatively poor eyesight, the trunk provides them with much of the information they need about their surroundings. From considerable distance, they can locate water sources or discern the reproductive status of other elephants or, given a favourable wind, detect the presence of man.

The strength of the trunk is formidable. Elephants routinely tear down thick branches and uproot shrubs with their trunks. They can lift great weights with them and hurl missiles. On the Rungwa River in Tanzania, Mike Carroll, an elephant hunter and artist, witnessed a cow seize from the water a large crocodile about fifteen feet long and weighing at least 680 kilos (1,500 pounds). With her trunk wrapped around its tail, she swung it high over her head, smashed it repeatedly to the ground and then flailed it against the bole of a tree for some five minutes.

Tusks, like trunks, serve principally as tools. They are used, together with trunks, to dig for water, for mineral salts and for roots; to prise bark from trees; and to lift weights. They act as trunk-rests. Just as humans are right-handed or left-handed, so elephants are right-tusked or left-tusked, using one in preference to the other. The tusk used most often—the master tusk—is shorter, worn down more, and more rounded at the tip.

Tusks are highly specialised incisor teeth in the upper jaw which first begin to protrude in calves about two years old. They continue to grow throughout an elephant's life, so that the older the elephant, the heavier its ivory. Dr Richard Laws, conducting research in Uganda, once calculated that if they did not break during a lifespan of sixty years, tusks would reach a length of sixteen feet in females and twenty feet in bulls. One quarter of the length of a tusk lies within the socket in the cranium, held there by a mass of tough, fibrous tissue.

The longest tusks recorded for a modern elephant came from one shot in the eastern Congo and obtained by the New York Zoological Society in 1907. The right tusk measured 11.4 feet on the outside curve, and the left tusk, 11 feet. Their combined weight was 133 kilos (293 pounds). The heaviest pair of tusks on record came from an elephant shot below Mount Kilimanjaro in 1897. Now housed in London's Natural History Museum, they weighed together 211 kilos (465 pounds), but measured only 10.2 feet and 10.5 feet.

Like other teeth, the tusk is composed mostly of dentine. In cross-section, however, what distinguishes it from other ivory is a pattern of lines that intersect each other to form diamond-shaped areas. This pattern is unique to elephant ivory. It is not found in the ivory of any other mammal, such as the hippopotamus, warthog, walrus, narwhal or sperm whale. It gives elephant ivory the distinctive lustre and adaptability that has made it fatefully so desirable to mankind.

The remainder of the teeth—molars used for grinding up vegetation—are also unique. Unlike other animals, elephants have a 'queue' of six sets of molar teeth in each jaw which move forward in horizontal progression as each successive set at the front wears out and breaks off, rather like a very slow production line. Only one set is fully in use at any one time. Elephants lose the first set at about two years old; the second at about six; the third at about fifteen; the fourth at about twenty-eight; the fifth at about forty-three. The sixth set appears at about thirty and comes into use in the early forties. This final set lasts for another twenty years. When it is worn down, elephants can no longer feed properly and die from malnutrition.

The ears are unusual too. They are particularly impressive when spread out in threat displays. They can measure up to six feet from top to bottom and more than five feet across. Ear flapping and folding are used as well to send out a variety of other signals to elephants within sight. The ears also serve as a highly efficient cooling apparatus. On the back of the ear is a network of blood vessels which enables hot arterial blood coming from the heart to cool by as much as 10° C before it filters back to the body. To activate the cooling system, elephants either extend their ears forward downwind to allow cooling air to blow across the backs of their ears or, if there is no wind, actively flap their ears like fans to generate a current of air.

The skull is massive, having grown during the course of evolution to be able to sustain the weight of the tusks and anchor the muscles of the trunk, but it is nevertheless much lighter than it appears. The brain case is surrounded not by solid bone but by a network of cavities, or sinuses, which provide the necessary skull volume and stability without excessive weight. The brain itself is located at the back of the skull, well away from the forehead. An adult elephant's brain weighs about five kilos (eleven pounds); though four times the bulk of a human brain, it is only one-tenth of the size relative to body weight.

Unlike most other mammals, elephants continue to grow in height long after sexual maturity. Females reach their maximum height by about twenty-five years of age, ten to fifteen years after puberty. Males continue to grow in height until about forty-five years of age. The largest elephant on record, shot in Angola in 1955 and now on display at the Smithsonian Museum, stood at thirteen feet two inches and weighed about ten tons.

To support their huge body, elephants have developed pillar-like legs. They normally move slowly, with short deliberate steps. They can cover ground more quickly with an extended walk, a shuffling run or a short sprint. But because they need to keep one foot on the ground at all times, they cannot run or sprint properly. Nor can they jump, trot, canter or gallop. With one leg out of action, they are practically immobile.

During the course of a normal day, elephants usually travel up to six miles, but they can cover as much as twenty miles at a stretch. Research on elephant movements in Samburu in northern Kenya has shown that they 'streak' between safe havens like game reserves and national parks, crossing intervening land at night to minimise the danger from hunters.

By outward appearance, their feet are broad, flat and solid-looking. But the bone structure is set at an angle, so that in fact they walk on tiptoe. Most of the body of the foot is composed of fatty, fibrous tissue, which acts as a cushion or shock absorber, enabling them to tread softly.

Their skin, marked by deep creases and wrinkles, is tough, thick and, to the human hand, feels rough. But it is surprisingly sensitive, both to touch and to temperature change. Elephants spend much time on skin care, bathing regularly and showering themselves with mud and dust to keep it free from parasites and disease. The folds and wrinkles in the skin also hold moisture to keep them cool.

When hot and tired from being chased, elephants try to cool themselves by reaching into their mouths with their trunks, regurgitating water from pouches halfway down their throats and showering their bodies with it.

Elephants have voracious appetites. They spend about three quarters of every day and night eating, drinking or moving in search of food and water. To keep up their bulk, adults need to consume around 150 kilos (350 pounds) of vegetation and up to 100 litres (175 pints) of water a day. They chew on almost anything: grass, leaves, herbs, creepers, twigs, bark, roots, sedges and aquatic plants. But they have a particular liking for ripe fruit and berries and crops such as maize and sugar. By standing on their hind legs, they can reach high up into trees for choice foliage. They also shake trees to bring down ripe fruit, and use their tusks to dig for roots and tubers.

They sometimes develop a passion for the fruits of certain trees—such as the marula tree or doum palms—which ferment in their stomach making them intoxicated. Hunters have recorded numerous instances of encountering inebriated elephants. In his book on South Africa's big game, published in 1875, W.H. Drummond described how elephants arrived in an area north of the Pongolo River just when the fruit of the umganu tree was ripening: 'This fruit is capable of being made into a strong intoxicating drink, and the elephants after eating it become quite tipsy, staggering about, playing huge antics, screaming so as to be heard miles off, and not seldom having tremendous fights.'

From his hunting expeditions in the Congo, Jean Pierre Hallet reported: 'I have more than once seen forest elephant, extremely drunk after gobbling up vast quantities of fermenting fruit, giggling with delight as they lobbed pieces of squishy fruit at each other's heads and derrières.'

Their feeding habits, however, are both extravagant and wasteful. They strip bark from trees for a few mouthfuls of food, leaving them to die. They push over whole trees just to get at leaves or fruit at the top. They even destroy giant baobabs, hundreds of years old. Only about half of their food intake is digested; the rest is excreted undigested. Though they distribute seeds for regeneration through their dung, they can leave behind woodlands that have been devastated.

Some become accustomed to foraging around human settlements. Rennie Bere, a game warden in Uganda, records how one medium-size bull known as the Lord Mayor took up residence at Paraa, the Murchison Falls Park headquarters, for five years in the 1950s.

He removed half the thatched roof from the warden's kitchen to get at a bunch of bananas and broke down a store to investigate a barrel of fermenting millet-beer; he regularly searched the staff village for scraps and turned out dustbins to see what he could find. He finished a whole sack of potatoes at a sitting; and one night he turned over a motor car whose surprised occupants were asleep inside. He learned how to turn on a tap when he wanted water (but not how to turn it off) and usually tore up the installation in disgust if the supply was cut off at the main. He developed a craving for human foodstuffs.

Researchers in Amboseli have found in elephant dung a host of items scavenged from tourist lodge refuse dumps: rubber gloves, condoms, ladies underwear, small bottles and batteries. But the liking for human foodstuffs can have fatal consequences. In her book about the Echo family in Amboseli, Cynthia Moss records the death of a forty-year-old cow, Emily, close to an unfenced refuse pit. Her stomach was found to contain bottle tops, glass, plastic, batteries and other items capable of perforating her intestines.

Like most of their activities, elephants synchronise their sleeping times. They fall into a deep sleep for between one and four hours each night, lying down on one side and often snoring contentedly. During the hottest hours of the day, they usually doze off while standing.

Mating occurs at any time of the day or night and at any time of the year, except when adverse environmental conditions such as drought cause them to defer breeding. Both male and female have unusual features about their sex organs. The male's penis is completely retractable; it nearly touches the ground when extended but otherwise disappears into a special pocket in the body wall. The testes do not descend into a scrotum, as with many mammals, but remain permanently located within the body, near the kidneys. The female's vulval opening lies not under the tail but between the hind legs, with a vaginal canal that follows a route unknown in any other terrestrial mammal.

Females have the longest pregnancy of any known mammal: twenty-two months. The foetus develops slowly, measuring less than four inches at four months and only twelve inches at one year. Even when it weighs 100 kilos (200 pounds) or so, females show almost no visible signs of swelling. In their lifetime, they may give birth to seven or more calves, continuing to breed into their fifties. Males also remain sexually active well into old age.

Elephants are renowned for having good memories and high intelligence. Aristotle in the fourth century BC referred to the elephant as 'the beast which passeth all others in wit and mind' and writers down the ages have been similarly impressed. Children are still taught that elephants never forget. Certainly, they have a remarkable ability for learning from past experience and quickly discern when something familiar to them is out of place. They are also quick to learn. In Kenya and elsewhere in Africa, when confronted with electric fences, they have soon discovered that while their bodies are vulnerable to electric shocks their tusks are not and can be used to break electric fence wires; they have also learned to drop logs on electric fence wires. When elephants in Addo National Park in South Africa were enclosed in 1953 by the world's first elephant-proof fence— made from train rails, set in concrete and connected with lift-cables—they tested it to the limit, pushing with their heads, sitting on the wires and rearing up on two legs in an attempt to pull it down.

In other signs of their intelligence, elephants use tree

branches to scratch their backs and legs in places where trunks and tails cannot reach, behaviour which biologists interpret as tool-using.

Their brain is highly convoluted, a feature they share with humans, the great apes and dolphins. Scientists trying to calculate an intelligence index for animals have awarded marks of 14 for wild boar; 48 for baboons; 104 for elephants; 121 for dolphins and 170 for humans.

Given good fortune, elephants live well into their sixties. They are subject to few diseases, although, like humans in old age, they are prone to cardiovascular diseases and arthritis. As their last set of molars wears down, they seek soft foods to chew, but eventually succumb to malnutrition. Old bulls die alone, but matriarchs remain with their family until the last.

# 21
## Rumbles

African elephants possess a rich and varied vocal repertoire audible to humans. While famous for their trumpets and screams, they use a wide range of other sounds to express themselves: a purring vibration seeming to denote pleasure; a soft, moaning squeal when experiencing loneliness. When within sight of each other, they also communicate through body language, by making subtle changes in the positions of their head, ears, trunk and tail. But the most intriguing aspect of elephant communication is the part beyond human hearing.

The complex world of elephant communication puzzled scientists for many years. Field biologists were familiar with what Rennie Bere, one of Uganda's most experienced game wardens, once called 'the interminable thunderous rumble of their bellies'. For centuries, hunters had ascribed the sound of 'tummy rumbling' to the elephant's noisy digestive system. Some scientists concurred. Writing about 'tummy rumbling' in her *Natural History of the African Elephant*, published in 1971, Sylvia Sikes, a scientist with years of field experience, reported: 'This sound is heard when an elephant, or a herd of elephants, is browsing peacefully and completely undisturbed. Presumably the outsize liquid pop-pop-pop-pop sound is that of the stomach contents, after a long drink, undergoing peristaltic churning, and is not respiratory in origin.'

Other scientists remained perplexed. In Manyara, Iain Douglas-Hamilton noted the ability of elephants to organise themselves silently without any visible or audible cue. In Amboseli, Cynthia Moss and Joyce Poole wondered how male and female elephants, living separately and far apart, managed to find each other during the brief and unpredictable phases when both were available for breeding. In Zimbabwe, Rowan Martin observed from his radio-tracking experiments that elephant herds sometimes coordinated their movements even when they were separated by miles of bush.

The explanation came in 1984 when Katy Payne, an acoustic biologist from Cornell University who had spent fifteen years studying the songs of humpback whales, decided to visit the elephant house in Washington Park Zoo in Portland, Oregon. Standing alongside an Asian elephant, Payne thought she detected a 'palpable throbbing like distant thunder', but she could hear no sound.

From her studies on whales, Payne was familiar with the way that these great mammals of the sea communicated by using infrasound, sound below the range of human hearing. But no land mammal was then known to use infrasound. Returning home, she wondered whether what she had experienced at the zoo was in fact infrasound: elephants communicating through sound too low for her to hear yet powerful enough to make the air throb.

With two colleagues, she set up a series of experiments at the zoo using equipment that could record and measure infrasound. When they played back the tape at ten times its usual speed, they were able to discern an array of elephant calls that none of them had heard during recording. At nearly three octaves too high, they sounded a little like the mooing of cows. The loudest calls coincided with a period when they had sensed throbbing in the air. Payne concluded: 'Two animals had been carrying on an extensive and animated conversation below the range of human hearing.'

In January 1985 Payne arrived in Amboseli to begin work with Joyce Poole on a study of elephant communication which opened an entirely new perspective on elephant life. Payne's subsequent studies took her to Namibia and Zimbabwe, while Poole remained in Amboseli spending six years unravelling many aspects of elephant language.

Elephant calls range from higher-frequency trumpets, screams, bellows and roars to lower-frequency growls and rumbles, like the rumbles once known as 'tummy rumbles'. While humans with acute hearing can pick up sounds with a frequency range as low as 20 hertz, many rumbles that

elephants make occur within frequencies between 14 and 35 hertz, which means a significant proportion remain outside normal human hearing. Forest elephants make calls as low as 5 hertz, two octaves below the lowest sounds that humans ordinarily hear. When using infrasound, elephants often combine low-frequency calls with high sound-pressure levels, as high as 115 decibels, close to the level of amplified rock music. Because low-frequency sounds travel for far greater distances than high-frequency sounds of the same pressure level and are less affected by woodland and other obstructions, this gives elephants an effective means of long-distance communication of up to six miles or more.

In Amboseli, Joyce Poole and her colleagues discovered that females are more talkative than males. In a study of twenty-six types of call made by adult elephants, nineteen were made only by females, three were made by adults of both sexes, and only four were made exclusively by males. Females use a range of calls to keep in contact both with family members and with other families for everyday purposes, while males communicate far less frequently, confining their calls mainly to matters of reproduction and male dominance. In research carried out in Amboseli in the 1990s, Karen McComb, a biologist from the University of Sussex, concluded that adult females were familiar with the calls of some fourteen different families, involving as many as 100 other adult females.

In all, Amboseli's researchers have identified more than fifty elephant calls, including thirty-five different rumbles, the most varied and complex part of the elephant's reper-toire. Among those in common use is the 'Let's go' rumble, made when a matriarch decides to make a move. Other calls advertise for mating partners: musth bulls use a musth rumble; estrous females use an estrous rumble to attract males. Females joined by a musth bull often answer his rumbles with a 'female chorus'. After copulation, females utter a sequence of rumbles, while family members gather round trumpeting and rumbling in mating pandemonium.

Poole has also identified a 'discussion' rumble. 'Elephants use those during times when they appear to be having a discussion, usually a disagreement, about the plan for the day's activities,' she writes in *Coming of Age with Elephants*. 'The apparently debating rumbles go back and forth between the individuals, sometimes for twenty minutes or more, very much in the cadences of a human conversation, but in slow motion.'

Some idea of the possible range of elephant communication is given by an incident that occurred in the Hwange area of Zimbabwe. Living on a private wildlife sanctuary adjacent to Hwange National Park were a group of about eighty elephants, a familiar sight to tourists at the lodge there. On the day that a culling operation started in the national park, ninety miles away, the elephants disappeared. They were found several days later in the opposite corner of the sanctuary as far away from the park boundary as they could get.

# 22
## *Death*

Elephants attach a particular significance to death. The loss of a family member, especially a matriarch, is marked by displays and rituals which some field biologists liken to a funeral. When attending dead companions, they move backwards, gently touching the body with their hind legs, circling, hovering, then touching it again. Elephants also perform burials, picking up branches, grass clumps and soil to cover the bodies not only of dead elephants, but of other dead animals and even dead humans they have killed. Long after death has occurred, they show an acute interest in elephant bones, in particular skulls and tusks, smelling them, turning them over, picking them up and carrying them some distance before dropping them.

An American wildlife biologist, Harvey Croze, witnessed the dying moments of an old matriarch surrounded by members of her family one afternoon in Serengeti. Lagging behind the family group, the matriarch stopped, swayed back and forth and fell to the ground. The others rushed back, trumpeting and rumbling, and clustered around her, putting their trunks in her mouth, pushing against her and trying to raise her. One young bull struggled repeatedly to lift her with his tusks, then stuffed vegetation in her mouth and finally tried to arouse her by mounting her. A calf knelt down and tried to suckle. Eventually the group started moving away, until only one cow and her calf were left. The cow did not face the body, but stood and reached back from time to time, touching it with a hind foot. Finally, as the other elephants called, slowly and seemingly reluctantly, she left the dead matriarch and walked away.

In Addo, Anthony Hall-Martin, a South African biologist who spent eight years on elephant research there, recorded what he described as a typical elephant funeral. Clustered around the body of a matriarch killed by a bull in musth, her entire family, including her young calf, touched her trunk and mouth and body with their trunks and tried to raise her. There was much rumbling. The calf cried and screamed loudly, but the others eventually fell silent. Then they began to throw soil and leaves on to the body, and tore up bushes and broke branches to cover it. For two days, they maintained their vigil, standing quietly near the body, wandering away on occasion to feed and drink, but always returning. Other elephants passing by came to sniff and touch the body, often just standing nearby, as though, said Hall-Martin, they were expressing sympathy.

In her book *Elephant Memories*, Cynthia Moss pieced together the circumstances surrounding the death of a young cow, Tina, as she tried to escape with members of her family from armed poachers who had attacked them outside Amboseli's boundaries. Badly wounded, groaning with pain, Tina fell behind her fleeing family. Her mother dropped back to run beside her, reaching over and touching her with her trunk. With blood pouring from her mouth, Tina collapsed. Her mother and an elderly sister made frantic efforts to hold her up, leaning on either side of her, but she slipped beneath them and, with a shudder, died. Other members of the family gathered round, trying to revive her. Using all her strength, her mother worked a tusk under her shoulder and started to lift her, but her tusk broke off with a sharp crack. Her family finally gave up, but did not leave and stood close to the body, touching it gently with their trunks and feet. Then they began to bury it with bushes, branches and earth. After standing vigil through the night, they slowly made their way at dawn to the safety of the park. Tina's mother was the last to leave, standing with her back to the body, gently touching it with her hind leg.

Joyce Poole tells the story of a cow in Amboseli who suddenly collapsed from snakebite and died on her own while family members moved on unaware. Following behind, four young bulls came across the body. The oldest male, Kasaine, wrapped his trunk around her trunk and tusks, made repeated efforts to raise her and spent an hour mounting her again and again, trying to get some response before walking away. The next day park rangers arrived to cut away the tusks, leaving her head mangled. Kasaine returned too, standing silently above the corpse, feeling the wounds gently with his trunk.

Various accounts have been recorded of elephants burying human victims. The German zoologist and conservationist Bernhard Grzimek collected four accounts, including one that occurred in a national park in the Congo. Ignoring warnings not to approach an elephant known to be dangerous, a tourist persisted in walking forward to take photographs. The elephant charged, caught the tourist with his trunk, knocked him to the ground, knelt on him and then drove a tusk through his body under the shoulder blade. When other members of the tourist's party returned to the scene, they found his body covered in plants.

A Kenya game warden, George Adamson, relates how an old Turkana woman, losing her way home one evening, lay down under a tree and fell asleep. She awoke a few hours later to find an elephant standing over her feeling her with its trunk. Paralysed with fear, she lay motionless. Other elephants arrived, and, trumpeting loudly, buried her under a pile of branches. The next morning, her faint cries for help were heard by a passing herdsman who released her.

In an incident on a Kenya ranch belonging to the conservationist Kuki Gallmann, a herdsman minding camels was knocked to the ground by a charging matriarch, breaking his leg. When he did not return, a search party set out the next day to look for him. They found him propped up against a tree, guarded by a lone female elephant. When they tried to frighten her away, she charged and chased them. Arriving at the scene, the ranch manager Colin Francombe tried to move her away with his vehicle but again she charged repeatedly. Assuming she was dangerous, he raised his rifle to shoot her. But the injured herdsman shouted out for him to stop. When she was finally driven off by gunfire, he described how the elephant, using her trunk and forefeet, had gently moved him several yards into the shade of a tree. Even when her own family had moved on, she had stood guard over him, throughout the night and into the next day, driving away a passing herd of buffalo.

As remarkable as their response to death is their fascination with the bones of the dead. Whenever encountering elephant bones, they stop to examine them, reaching forward tentatively at first to smell them, then picking them up. Tusks excite particular interest and are sometimes passed from elephant to elephant.

In Amboseli, Cynthia Moss witnessed an occasion when members of the Echo family approached the bare, bleached bones of a cow named Emily who had once been a prominent family member.

The first to reach the bones was a young female calf, Eleanor, and right behind her were Erin and Edgar. These three immediately stopped and cautiously reached their trunks out. They stepped closer and very gently began to touch the remains with the tips of their trunks, first light taps, smelling and feeling, then strokes around and along the larger bones. Eudora and Elspeth, Emily's daughter and granddaughter, pushed through and began to examine the bones, and soon after Echo and her two daughters arrived. All the elephants were now quiet and there was a palpable tension among them. Eudora concentrated on Emily's skull, caressing the smooth cranium and slipping her trunk into the hollows in the skull. Echo was feeling the lower jaw, running her trunk along the teeth—the area used in greeting when elephants place their trunks in each other's mouths. The younger animals were picking up the smaller bones and placing them in their mouths, before dropping them again. The spell was broken when one-year-old Edgar irreverently began tossing ribs in the air. After a few more minutes they all began to move off, some carrying bones with them, either in the trunk or wedged between the trunk and tusk.

Once Moss collected the jawbone of an old adult cow that had died of natural causes, taking it back to her camp to determine her age. Three days later, the cow's family passed through the camp, smelled the jaw and made a detour to inspect it before moving on. But one young elephant stayed behind for long after the others had gone, repeatedly feeling and stroking the jaw and turning it with his feet and trunk. It was the dead cow's seven-year-old calf.

George Adamson records how he once shot a bull

elephant among a group that had been raiding a government official's garden in northern Kenya. After allowing local Turkana tribesmen to carve up the meat, he dragged the carcass about half a mile away. That night, elephants visited the body, picked up a shoulder blade and a leg bone and returned them to the exact spot where the elephant had been shot.

An even stranger incident was reported to have occurred in Murchison Falls Park in Uganda, where a group of scientists and park officials were working on a culling programme. As well as using elephants for meat, they stored their ears and feet in a shed to sell later for making handbags and umbrella stands. One night a group of elephants broke into the shed and buried the ears and feet.

Field biologists often debate the extent to which elephants experience emotions similar to grief. Like other animals, they display sorrow through their movements, postures and actions. But watching them stand disconsolately by the body of a dead calf or some other family member, their trunks hanging limply down to the ground, always reluctant to leave, gives the impression that deeper feelings are involved. Certainly, the death of a prominent family member, like a matriarch, can have a profound effect on the rest of the family, leading to the collapse of their normal routine.

Though no one has ever discovered an elephants' graveyard, the myth of their existence still survives. On occasion, large concentrations of elephant bones have indeed been found. In the days when African hunters used fire-rings, elephants perished en masse in the blaze, leaving behind piles of bones which travellers are said to have mistaken as evidence of graveyards. One of the last recorded incidents

of fire-rings occurred in the southern Sudan in 1927 when 277 elephants were burned to death, certainly enough to fill a cemetery. Elephants dying in large numbers of disease or drought may have left the same impression. Old elephants on their last molars sometimes retire to areas where soft vegetation is available, also leading to an accumulation of bones. But none of this explains why belief in the existence of elephant graveyards remains so prevalent.

The myth survives most probably because of the enduring popularity of *Arabian Nights*, which includes the story of how Sindbad the Sailor, after hunting elephants for his master, was taken by a herd to a hillside entirely covered with bones and tusks. 'This sight filled my mind with a variety of reflections,' says Sindbad. 'I admired the instinct of these animals, and did not doubt that this was their cemetery or place of burial, and that they had brought me hither to show it to me, that I might desist from destroying them, as I did merely for the sake of possessing their teeth.'

Though more than 1,000 years old, the story carries a message of particular relevance for modern times.

# 23
## The Culling Option

During the 1960s, as elephant populations in national parks and game reserves were swelled by ever growing numbers seeking refuge from human encroachment, the problems of elephant management in some areas became acute. The difficulty was illustrated most vividly in Uganda. In a paper published in 1962, Irven Buss and Allan Brooks showed that between 1929 and 1959 the area occupied by elephants had fallen from 70 percent of all land to only 17 percent, while the human population had risen from 3.5 million to 5.5 million. Year by year, more and more elephants were being 'compressed' further in their safe havens by immigration. The result in places such as the Murchison Falls National Park was that areas of thick woodland had been converted into open grassland, putting at risk not only elephant populations but a variety of other wildlife. A large proportion of bird life had already disappeared; so had the chimpanzee and the forest hog. In optimum conditions, elephant populations were capable of rapid growth of 6 percent a year, doubling their numbers in fifteen to twenty years. But now they no longer had the freedom to roam outside the parks, a management crisis was in the making.

The 'elephant problem', as it was called, and the way to tackle it generated intense controversy. One school of thought, led by the scientist Richard Laws, argued that the only solution lay in culling elephants in sufficient numbers to restore the balance between elephants and their habitat. Since national parks and reserves in the first place were man-made areas with artificial boundaries, effective management there was necessary to ensure a balance was maintained. Unless elephant numbers in national parks and reserves were reduced to counteract excessive immigration, the whole ecosystem might crash, causing the extinction not only of elephants but of other species.

The other school of thought insisted that culling would subvert the essential purpose of national parks and conservation. The best option, they argued, was not to intervene. If elephants were left alone, they would eventually reach a balance with their environment. The decline of woodlands was not necessarily irreversible but part of a long-term cycle that elephants had always influenced.

The solution chosen in Uganda was to cull. Between 1965 and 1967 some 2,000 elephants out of a herd of 14,000 in Murchison Falls National Park were shot. Whole families were wiped out to prevent panic and fear spreading through the park. As they grouped tightly bunched in defensive circles, with mothers facing outwards and calves hidden behind, professional hunters with semiautomatic weapons opened fire first on older females, then finished off the rest as they milled about. In most cases, a family group of twelve was dead within a minute.

The cull was used by scientists to enhance their research data. Each corpse was painstakingly dissected, weighed and measured. The results, published by Laws in scientific journals, became the standard reference work on elephant reproduction and population dynamics. But the use of culling nevertheless plunged the wildlife community into bitter strife which lasted for decades.

A similar problem occurred in Kenya's Tsavo National Park. Once covered in *nyika* thornbush, much of it had been reduced within a period of twenty years to a landscape of open savannah and smashed trees, reminiscent in parts of the Somme battlefields of the First World War. Giant baobab trees, hundreds of years old, had been felled by the score.

The warden of Tsavo, David Sheldrick, had devoted twenty years to developing and protecting the park, making it the most famous in Kenya. During the 1950s, with a field force of only fifty men to cover an area of 5,500 square miles, he had fought a prolonged campaign against Waliangulu elephant hunters armed with longbows and arrows tipped with deadly poison from the *Acokanthera* plant, which dropped elephants in their tracks within minutes. It was largely as a result of Sheldrick's success against the Waliangulu that Tsavo's elephants were thriving in such numbers.

To help him assess the scale of the problem, Sheldrick arranged for Richard Laws, fresh from his assignment in Uganda, to carry out a scientific cull of 300 elephants in order to obtain a profile of their ages and reproductive trends, as he had done in Murchison Falls National Park. Laws's findings, which he announced in 1969, provoked uproar. He warned that unless action was taken, Tsavo would become a desert. He recommended a cull of 3,000 elephants for further research purposes.

Sheldrick balked at the idea. Though he had once thought that culling might be unavoidable, he now took the view that minimal interference was preferable even if it meant that large numbers might die of starvation. The park authorities supported Sheldrick and Laws resigned.

In the event, Tsavo was overtaken by a severe drought in 1970–71 which led to the death of some 10,000 elephants, a quarter of the population. Both sides in the culling controversy claimed vindication. The pro-culling camp argued that the high level of mortality was due to the effects of overpopulation which culling would have averted. The anti-culling camp maintained that the drought showed how natural events kept the population in balance with the environment, thus justifying a policy of non-intervention.

Elsewhere in Africa culling programmes were readily adopted as the answer to overpopulation. In Zambia's Luangwa Valley some 1,500 elephants were killed between 1965 and 1969. A permanent abattoir was built on the banks of the Luangwa River to process the carcasses. Initially, elephants were shot with a dart giving them a drug overdose which avoided the disturbance caused by rifle shooting. But it also meant that whole families could not be eliminated at once, thereby spreading fear and confusion in the park. Moreover, the drug worked by paralysing the respiratory muscles, causing the elephant to die from suffocation, a distressing form of death. In 1968 the culling teams reverted to rifle shooting.

In Rhodesia (Zimbabwe), where the elephant population had climbed from an estimated 4,000 in 1900 to more than 60,000, culling became regular practice, even in areas where no severe elephant damage had occurred, in order to keep elephant populations at predetermined levels. In the Wankie (Hwange) National Park, initially some 3,000 elephants were killed between 1971 and 1974, after which the park authorities set a target of culling between 3 and 4 percent of the population to maintain numbers at around the 13,000 level. Other culling programmes were carried out

regularly in Sengwa, Matusadona and Gonarezhou. Helicopters or light aircraft were used to herd families towards waiting marksmen. Only young calves aged between one and three years were spared as they milled about their dead relatives, to be sent for sale to zoos and safari parks abroad.

The culling business in Zimbabwe provided a useful source of income. Elephant meat and fat were sold locally; tusks were sent off to ivory auctions; hides were used for briefcases, shoes, bags and other leather products; and feet were turned into foot-stools and umbrella stands; even the feet of baby elephants were made into pencil holders and cigar containers.

Indeed, so important did the income factor become that

it eventually overtook the ecological reasons for culling. Wildlife officials, such as Rowan Martin, argued that with 17,000 square miles of national park land to protect, costing US$500 a square mile, elephant culls were an essential means of raising money. Sustainable utilisation became a central theme of government policy. Elephants were expected to pay their way. Between 1981 and 1988 Zimbabwe culled nearly 25,000 elephants, earning more than US$13 million from elephant products.

The statistics of culling were kept in meticulous detail. Measurements were taken of each dead elephant's body, of its tusks, of each unborn foetus, of each calf captured alive. Nothing was ever recorded beyond physical data. But a brief glimpse into the lives of culled elephants came from a cull in Sengwa Wildlife Research Area in 1991 in which 249 elephants—one third of the population—were killed. Among the victims were five elephants once well known to Katy Payne from her studies there in elephant communication. To mark their passing, she listed them in her book, *Silent Thunder*—Jabula, Friday The Thirteenth, Munyama, Miss Piggy and Runyanga—giving what biographical detail she knew.

Jabula was the most familiar to her. An old matriarch, the mother of seven calves, she had been given radio collars in 1980, 1982, 1984 and 1986, enabling researchers to keep a regular check on her movements. Payne had located her on hundreds of occasions. Most of the time she spent in the company of her family of about twelve close to a permanent muddy water hole that elephants had gouged out of clay in the Lutope flood plain. She was in a nearby gully in September 1991 when she was killed, along with twelve other elephants. Two other elephant groups, numbering in all fourteen, were killed at the pool; one group of seven consisted entirely of young males.

On a subsequent visit to Sengwa, Payne tracked down the records of the cull. She found the elephants duly listed, their size measured according to a uniform standard called the 'adult equivalent' used for the purpose of estimating their value. The entry for the death of Jabula's extended family read as follows:

| | | | |
|---|---|---|---|
| Adult equivalents shot: | 20.911 @ $1,330 | = | $27,812 |
| Adult equivalents caught: | 0.587 @ $1,330 | = | $ 781 |
| Total adult equivalents: | 21.498 @ $1,330 | = | $28,592 |
| Total ivory: | 146.088 @ $ 200 | = | $29,218 |

South Africa pursued a similar culling strategy. From the score or so that had survived the mass slaughter in the lowveld in the nineteenth century, the elephant population in Kruger National Park by the 1960s had reached nearly 9,000. The park authorities decided that the optimum level there was about 7,500 and began regular culling in 1968, killing up to 500 each year for the next three decades. The carcasses were taken to a processing factory at Skukuza in the middle of the park, a mile away from the main tourist centre, cut up and turned into canned meats, dried biltong, leather for boots and belts, and bones for cattle feed.

Young orphaned elephants were sometimes spared for relocation, but adolescents growing up without adult discipline soon became known for wayward and aggressive behaviour. During the 1980s a group of bull elephants aged between eight and ten years old, survivors of a culling operation in Kruger, were transferred to a new national park in Pilanesberg in western Transvaal, where there were no adult males. They entered prolonged musth periods at an early age, fought among themselves and attacked other wildlife. Between 1992 and 1997 they killed more than forty white rhinoceroses. Two tourists were also killed and a number of vehicles damaged. The killing stopped in 1998 after six older male elephants were shipped in from Kruger, establishing a new male hierarchy and keeping the adolescents in check.

In a change of policy in 1995, the park authorities in Kruger suspended culling operations and turned their attention to research on other ways of controlling the population level. The most innovative idea involved contraceptive drugs for elephants. Scientists at the University of Georgia developed a contraceptive vaccine that causes the elephant's immune system to produce antibodies which prevent fertilisation. When administered to elephants during trials in Kruger, the new birth-control drugs reduced elephant pregnancies by up to 70 percent.

Nowhere was the plight of elephants facing human encroachment more poignant than in Rwanda, a small, heavily populated country in central Africa. By 1973 only two elephant herds, numbering in all about 140, survived amid a sea of humanity. Week by week, as farmers expanded into new land, their ranges shrank. The elephants in turn took to raiding crops, devastating the fields of an experimental agricultural station growing rare strains of banana.

After two years of debate, the Rwandan authorities agreed to drastic action, deciding that all elephants should be eliminated except youngsters aged between one and ten years. Those under one year would also be killed as they would be too difficult to rear. The remaining youngsters would be transferred to the Akagera National Park, fifty miles away, where there were large herds of buffalo and zebra, but no elephants.

The elephants were duly herded by helicopter towards the guns of professional hunters. More than 100 died. But twenty-six youngsters—twelve males and fourteen females —were captured. The smaller ones were airlifted out by helicopter; the others followed by road. Once in Akagera, they settled down into a stable social group; and in 1983, three calves were born forming the nucleus of a new population.

# 24
## *Ivory Wars*

Just when scientists were beginning to understand the true nature of elephants, a new onslaught was unleashed against elephant populations across Africa. Starting in the early 1970s, it lasted for nearly two decades and cost the lives of hundreds of thousands of elephants, bringing some populations close to extinction. It was fuelled mainly by an insatiable demand for ivory from newly prosperous Asian countries, such as Japan, causing ivory prices to soar. After remaining stable throughout the 1960s, the price of ivory jumped from US$5.50 per kilo in 1969 to $7.50 in 1970; to $74 in 1978; to $120 in 1987; and to $300 in 1989. Elephants in the bush were suddenly worth small fortunes, not just to poachers but to a host of middlemen—smugglers, businessmen, corrupt officials and greedy politicians—who settled over the trade like flies. 'Safe' populations in national parks were no longer safe. Even game departments joined the onslaught. Adding to the toll was a growing proliferation of wars and civil conflicts in Africa that made automatic weapons readily available to poaching gangs. International efforts to control the trade proved worthless. Meanwhile, wildlife scientists and officials fell into a protracted dispute about the gravity of the problem. In wildlife circles, it became known as the Ivory Wars.

The first signs of the onslaught appeared in Kenya in 1971 when gangs of hunters, armed with automatic weapons, moved into Tsavo. The warden, David Sheldrick, was at first successful in holding the line against them. But simultaneously there was mounting evidence of corruption at senior levels of the Kenya Game Department, which was responsible for the control of problem animals outside national parks and for granting hunting licences. With the support of leading politicians, the Game Department issued large numbers of 'Collector's Letters' which gave holders authority to handle ivory said to have been 'found' on dead elephants. Among the politicians involved in the ivory trade were members of President Kenyatta's family. Large consignments of ivory were flown out of the country, often with false documentation. Despite clear evidence of poaching rackets and illegal trading, no cases were brought to court.

In response to international criticism, the Game Department agreed to hold a seminar in Nairobi for biologists, wardens, conservationists and professional hunters to discuss the problem. On the initiative of Peter Jarman, a research biologist employed by the department, the participants were invited to contribute data towards an estimate of the elephant population in Kenya. The figure they arrived at was 167,000, of which 49,000 were estimated to be based in national parks and game reserves.

The seminar heard forthright views on the level of government corruption. A key figure in the proceedings was Ian Parker, a former game warden who had set up a private culling business, winning his first contract to cull 2,000 elephants in Murchison Falls National Park in the mid-1960s. He had also made a detailed study of the East African ivory trade. Parker left his audience in no doubt about the involvement of the Game Department and senior politicians in the poaching system. But he explained poaching losses and the decline of elephant numbers generally as the outcome of human population growth and shrinking elephant ranges, not as a result of the demand for ivory or the rising price for it. Because of his grasp of detail, Parker's views held sway for many years to come.

Peter Jarman was no less forthright in setting out his views about corruption and he submitted a confidential report to the senior game warden. Within two weeks he was informed his contract would not be renewed.

The onslaught gathered momentum, overwhelming whatever defences Sheldrick could muster in Tsavo. His small field force arrested wave after wave of poachers, but steadily lost ground. In 1976 he estimated that in the previous two years Tsavo lost some 15,000 elephants, nearly half of the park's population.

Worse was to come. Ignoring all protests, the government decided to merge the management of the national parks with the Game Department. Ostensibly, the purpose was to streamline two organisations dealing with wildlife. But for the national parks organisation, the merger seemed a recipe for disaster. With an independent board of trustees, it had made determined efforts to protect the parks. The Game Department was notorious for inefficiency and corruption.

In 1976, after submitting a confidential report on the scale of poaching, David Sheldrick was removed from Tsavo. The warden who replaced him was ineffective and resigned after six months. The next warden was posted from the Game Department. From that point, Tsavo became an open killing field. Even after the government announced a ban on hunting and on trade in ivory, it made little difference. By 1979 Tsavo's elephant population had fallen to about 11,000. Other areas suffered a similar fate. Kenya by 1979 had lost half of its population—some 70,000 elephants.

The collapse of Uganda's elephant population was even swifter. After General Amin took power in 1971, Ugandan troops plundered national parks at will. When wildlife biologists flew into Rwenzori National Park (formerly Queen Elizabeth National Park) in 1976 to attend a seminar at the tourist lodge on the Mweya peninsula, from the air they counted more carcasses of dead elephants than living ones. Two Cambridge scientists, Keith Eltringham and Robert Malpas, who had continued working in Uganda despite Amin's chaotic rule, reported to the seminar that the population of Murchison Falls National Park had fallen from 14,300 in 1973 to 2,250 in 1975. The plains there were littered with bones. Their estimate for the national population was that it had plummeted from 60,000 to 6,000.

International efforts to control the ivory trade came to nothing. In 1976, at the inaugural Conference of the Parties to the Convention on International Trade in Endangered Species of Flora and Fauna (CITES), the African elephant was listed on Appendix II. This meant that ivory exports required an export permit, but the system was open to any amount of abuse. Although it was recognised almost immediately that the machinery of CITES Appendix II was inadequate to the task of regulating the ivory trade, no further measures were taken for another ten years.

Nor were scientists agreed among themselves about the scale of the problem. Prominent figures such as Richard Laws still argued, as he had done in the 1960s, that the greatest threat facing elephants in East Africa was over-population in protected areas for which the solution was culling. Laws wrote in 1978:

For decades, the political authorities in East Africa have explained away their own very parochial failures by attributing the decline of elephant populations to poaching in the mistaken belief that the problem will evaporate once poaching is controlled. But the problem will not evaporate. The anaesthetic that envelops it is wearing thin, and soon it will come to the full consciousness of all, that the increasing concentration of any species in arbitrarily restricted areas will result in overpopulation, habitat destruction, and local—perhaps even wholesale—extinction of that species.

The question is this. Should this species [the elephant] be allowed to find its own level in relentless competition with others for the world's limited land and resources, or should this species be brought into balance with the environment and other species that inhabit it?

The case for culling in overpopulated parks preoccupied other reputable wildlife experts, including Peter Scott, founder of the World Wildlife Fund. 'As an example of the scale of the problem,' wrote Scott, 'it is estimated that there are about one hundred thousand elephants in the Luangwa Valley in Zambia, and that for the habitat to recover, that number must be reduced to twenty-five thousand. The idea that seventy-five thousand elephants must be culled will be very difficult for most people to accept. Nevertheless it seems necessary for WWF to begin at once to explain to its constituency this unhappy paradox of world depletion and local over-abundance.'

Little account was taken of whether the poaching

epidemic that had struck Kenya and Uganda would spread elsewhere in Africa. No one was sure, in fact, what was happening across great swathes of Africa either to elephant populations or to the ivory trade.

Two major reports published in 1979 were swiftly engulfed in controversy. The first was by the Manyara biologist, Iain Douglas-Hamilton, who had spent three years engaged in a pan-African survey of elephant populations, a project financed by the World Wildlife Fund, the New York Zoological Society and the International Union for the Conservation of Nature (IUCN), a consortium of government and non-governmental conservation agencies. From his investigations on the ground in Zaire (Congo-Kinshasa) and the Central African Republic, both with large elephant populations, Douglas-Hamilton was convinced that government elites, in league with European traders, were engaged in the wholesale export of ivory. Illegal trading was also widespread throughout West Africa. Using evidence from questionnaires circulated to wardens, conservationists and scientists throughout Africa, he gathered together counts and estimates of elephant populations, arriving at an overall figure of 1.3 million, seemingly a safe enough figure. But one respondent after another also warned of a sharp fall in elephant numbers, precipitated by the soaring price of ivory. Only a few countries in southern Africa—Zimbabwe, Zambia, Botswana and South Africa—reported stable or rising populations.

Fearing a major crisis, Douglas-Hamilton drew up an African Elephant Action Plan. He called for measures both to reinforce protected areas and to tackle the burgeoning illegal ivory trade. 'The only way to counter the wider ivory trade is for united international action. CITES provides the framework, but, in addition, there is need for fully coor-dinated police action against the illegal ivory traders. It should mean throwing open all the accounts, documents and trade secrets of the companies dealing in ivory.' He dismissed the idea of self-regulation of the ivory trade and recommended total bans in selected countries.

The second report was an investigation into the inter-national ivory trade by Ian Parker. Parker had been recommended for the assignment by Douglas-Hamilton during discussions with the US Fish and Wildlife Service which financed the project. He had displayed both courage and skill in exposing corruption in Kenya and he was widely acknowledged as an expert in the field. But he himself was to become an ivory trader, committed to the trade.

Parker provided a mass of new detail about the workings of the ivory trade. He confirmed that the trade had reached levels not seen since the early twentieth century. He calcu-lated that between 1973 and 1978 about 1,000 tons of ivory a year had left Africa. Ivory, he suggested, was seen as a source of wealth which, at times of political instability, could be exported, like gold or diamonds. But he concluded that the ivory trade was not excessive; that it fell well within sustainable limits; and that it did not threaten the elephant's survival. If anything, he said, the threat came from population growth and loss of habitat.

Parker went on to accuse conservationists of using false statistics to conjure up an elephant crisis to enable them to raise funds for their own purposes. He argued that even on the basis of evidence contained in the Douglas-Hamilton report, the estimate of elephants should have been 2.5 mil-lion, not 1.3 million. Furthermore, he pointed out, the report had classified more than 90 percent of the population either as safe or of unknown status, which could hardly be said to constitute a crisis.

The arguments came to a head in 1981 at a meeting in Hwange in Zimbabwe of the IUCN's African Elephant Specialist Group, a technical committee consisting of representatives from across Africa. Parker's views prevailed. His figures were regarded as the most authoritative available. Douglas-Hamilton's figures were dismissed as guesses. Only in two countries—Kenya and Uganda—was there said to be evidence of a steep decline in numbers. As for the rest, it was agreed that there was no cause for alarm. As hosts of the meeting, Zimbabwe's representatives impressed on other delegates the benefits of its culling programme, stressing the importance of utilising elephants as an economic resource.

At a meeting of the specialist group the following year, Parker and the ivory lobby once again prevailed. The volume of ivory traffic, he argued, was driven not by the rising price of ivory but by human expansion into elephant ranges. Once again, Douglas-Hamilton's views were given short shrift.

Yet Parker's position was fundamentally flawed. He had underestimated the real impact of the ivory trade by failing to take account of the way in which an ever-increasing toll on elephant populations was needed to sustain the same volume of trade. Since 1979, the Wildlife Trade Monitoring Unit (WTMU) had begun compiling trade data as part of its effort to monitor enforcement of the CITES Appendix II agreement. Its figures showed that the mean tusk weight of ivory imported into Japan dropped from 16.2 kilos (35.6 pounds) in 1979 to 9.6 kilos (21.1 pounds) in 1982. What had happened was that poachers had started out by seeking elephants with large tusks—mainly mature bulls—which fetched a higher price. When mature bulls were no longer to be found, they began to target cows with a much lower average tusk weight. By 1987 the trend had become even clearer. The average tusk size exported was down to 4.7 kilos (10.3 pounds). Whereas in 1979 one ton of ivory had represented about 54 dead elephants, by 1987 it represented a minimum of about 113 dead elephants. Because so many were adult females, the losses went even higher. Studies in Amboseli showed that for every adult female killed at least one immature elephant would die. A calf younger than two years stood no chance of survival; a calf orphaned between two and five years old had a 30 percent chance of survival; and a calf aged between six and ten years old had a 48 percent chance of survival. Thus, added to the number killed by poachers were an estimated 55 calves with no ivory who died as orphans, bringing the figure to nearly 170 elephants per ton of ivory exported. An additional impact was the disruption to breeding patterns. In heavily poached areas in Kenya and Uganda, the loss of so many mature bulls led to seriously skewed populations.

When wildlife scientists drew up computer models to test the impact of poaching on elephant populations, using all the information available from trade statistics and elephant statistics, including age structures, reproduction, mortality rates and growth rates of tusks, their conclusions were unavoidable. 'The herds kept crashing to extinction,' said Jonah Western, chairman of the specialist group who commissioned the first studies in 1982.

One vital missing factor was the size of the forest elephant population in countries such as the Congo, Gabon and the Central African Republic. Parker asserted that a significant proportion of ivory exports consisted of the distinctive tusks that came from forest elephants. The high volume of forest elephant ivory exports, together with the large average size of tusks exported, he claimed, were evidence of a huge population. He put the figure at between 2 and 3 million.

Yet no one had any clear picture of the forest elephant population. Living beneath the canopy of the uncharted central African rainforest, an area stretching over 1 million square miles, they remained hidden from aerial surveys. Nor were they easily accessible from the few roads that penetrated the region. To tackle the problem, Jonah Western commissioned Richard Barnes, a British biologist renowned for his tenacity, to work out a system for producing a reliable estimate. For several months, Barnes attempted a direct count of elephants in Gabon from the ground, but never saw more than a handful. Instead of counting elephants, therefore, he decided to count elephant dung. It was a time-consuming business. To make the survey scientifically credible, he had to cut straight lines, or transects, through the forest, a dozen miles in length, counting elephant dung along the way, crossing marshes and rivers and whatever else lay in his path.

The poaching epidemic, meanwhile, spread further and further. Elephant herds in national parks such as Selous in Tanzania and Luangwa in Zambia were decimated. Researchers returning to Uganda's Murchison Falls National Park, following the chaotic aftermath of General Amin's overthrow by Tanzanian forces, found only 400 elephants left, most in a single herd that stampeded across the open grasslands at the first sound of an engine. From Sudan,

Somalia, Congo and the Central African Republic, ivory exports flowed in an endless stream, covered by false permits.

The main demand for ivory came from Asia. Hong Kong and Japan together accounted for about 75 percent of total world imports. In the early 1980s, Hong Kong's imports exceeded 500 tons annually, more than 50 percent of Africa's exports. About one third was re-exported as raw ivory to other Asian countries: Japan, China and India. The remainder was consumed by Hong Kong's own carving industry which turned out a variety of sculptures, jewellery and other ivory products for sale to markets around the world, including the United States, Europe and Japan. Little remained in Hong Kong itself.

Family firms based in Hong Kong controlled much of the trade in Africa, paying a network of government ministers, diplomats, police and customs officials, army officers and game department personnel to ensure the ivory kept flowing. Prominent traders such as the Poon brothers, the Lai family and K.T. Wang made huge fortunes from the ivory business.

Japan was by far the largest consumer of ivory. Fuelled by rapid post-war economic prosperity, its imports of raw ivory increased from under 100 tons in the early 1950s to 600 tons in the late 1970s. By the mid-1980s, Japan accounted for nearly 40 percent of all worked ivory produced throughout the world. Much of it was consumed domestically. One item alone—*hanko*, personal signature seals used for everyday purposes such as business and legal contracts, bank cheques, and certificates of birth, marriage and death—swallowed 65 percent of Japan's raw ivory imports. Once manufactured predominantly from wood and horn, by the 1970s *hanko* made from ivory had become a highly fashionable product. Ever more elaborate versions were in demand. By the 1980s, about 25 percent of the world's consumption of raw ivory went into the production of signature seals for Japan.

China's ivory carving industry came third in the league after Japan and Hong Kong; it exported its products mainly to Hong Kong for onward sale. Other rapidly developing countries such as Taiwan and Korea joined the trade. Singapore became a major staging post, acquiring large stockpiles.

Further demand came from Europe and the United States. Though neither possessed significant ivory-carving industries any longer, both imported large quantities of ivory products from countries such as Hong Kong and China, keeping the trade buoyant. Europe imported some 18 percent of the world's worked ivory; the United States, about 16 percent.

As the scramble for ivory became ever more rapacious, CITES lumbered into action once more. Acknowledging that more needed to be done to protect the African elephant than listing it on Appendix II, a CITES conference in 1985, attended by representatives from nearly 100 member states, authorised the establishment of a quota system intended to bring the ivory trade under closer control. The quota system required each ivory-producing state that was a signatory to the CITES treaty to set an annual quota on the amount of ivory it would export based on a management programme for the sustainable utilisation of its elephant stock. Exporting states were required to notify the CITES Secretariat of any ivory consignment exported under its quota and to mark each tusk indelibly to identify its quota number.

Importing states were required to accept only ivory shipments covered by valid documents.

The quota system was a total failure. Bureaucratic rules were no match for the legions of armed gangs, corrupt officials, greedy politicians and rich traders at work in the trade. The CITES Secretariat had no means of enforcing the rules. Its own 'ivory unit' was so short of funds that it depended on 'donations' from wealthy ivory merchants to keep it functioning. Between 1985 and 1989 two-thirds of its budget was provided by ivory dealers. African governments, meanwhile, set whatever quota they liked. When they found themselves with more ivory than expected, they simply ignored their quota, exporting at will. Permits were forged or issued fraudulently just as regularly as before.

There were other massive loopholes to the system. Worked ivory was considered too difficult to monitor, so it was excluded. This meant that all that ivory traders had to do to avoid controls on raw ivory was to convert it to 'worked' ivory. In one case, Hong Kong traders sent 67 carvers and 150 labourers to Dubai in the United Arab Emirates to set up two carving factories. They purchased poached ivory at a reduced price, had it carved in Dubai sufficiently to pass as worked ivory, imported it legally to Hong Kong and sold it at a much higher price on the world market. Hong Kong companies acquired vast stockpiles of ivory by such means. Traders also avoided controls simply by establishing entrepôts in countries such as Taiwan that were not signatories to the CITES treaty.

The CITES Secretariat itself facilitated the legalisation of large parts of the illegal traffic in ivory. To encourage non-member states to sign up to the treaty, it agreed to wipe the slate clean to enable all tusks in their possession to be registered for sale. The result was that large stockpiles of suspect ivory were made legal with CITES approval, doubling their value overnight. Somalia's export quota for 1986 was 17,000 tusks, more than what was left in its entire national herd, providing a handy bonus for ruling politicians who were directly involved in the poaching business. Singapore, after acceding to the CITES treaty in 1986, was allowed to register its stockpile of 300 tons of ivory, making instant fortunes for the traders who had acquired it. The Poon brothers alone made nearly US$8 million overnight.

The most blatant example was Burundi, a small land-locked central African state. The only elephant it possessed was a lonely figure in the national zoo. But it was the hub of a smuggling network drawing in ivory from the Congo, Tanzania, Zambia and Mozambique. According to Burundi's customs statistics, it exported 1,300 tons of ivory between 1965 and 1986. In 1986 CITES 'legalised' 89 tons of ivory held by Burundi on condition that no more ivory was smuggled through the country. Despite the agreement, Burundi exported an additional 110 tons between November 1986 and October 1987.

The quota system was in effect being used to legitimise the illegal killing of tens of thousands of elephants. But officials at the CITES Secretariat argued that, given time, it would work. It was better to encourage the trade to develop legally, they said, than to drive it underground, which would result in even higher prices for ivory and greater risk to elephant populations.

All the arguments over the ivory trade raged anew at a meeting of the Elephant Specialist Group held in Nyeri, Kenya in 1987. A strong case for supporting the trade was made by Rowan Martin, a respected scientist from

Zimbabwe where elephant populations were rising. Martin had been one of the principal architects of the quota system. He argued that high prices for ivory, far from threatening elephant populations, provided the means to protect them. What was needed in Africa, he said, was an effective system of protection and utilisation, like the one that Zimbabwe operated.

But the evidence about the impact of the ivory trade on elephant populations across much of Africa was by now irrefutable. One survey after another carried out by scientists recorded a drastic fall in numbers. From the rainforests of central Africa, Richard Barnes reported a similar decline. Trade statistics showed that in a period of five years tusk weights had halved. Computer models predicted collapse. By the end of the meeting there was general consensus that the ivory trade represented a primary threat to elephant survival.

When the group's findings were reported to the next CITES conference in Ottawa in August 1987, they were duly adopted. But CITES did nothing about the trade.

The reason was that the prevailing policy within the world's conservation establishment—including the IUCN, the World Wildlife Fund and CITES—was to support sustainable utilisation. African governments were equally keen on the idea, hoping to raise substantial funds. In the case of elephants, the example that all cited was the countries of southern Africa where the largest remaining herds of savannah elephants lived, largely free from poaching. To try to ensure that their elephant herds maintained a balance with their environment, both South Africa and Zimbabwe had developed an infrastructure to cull and process thousands of elephants each year, using the revenue from elephant products to help maintain national parks. South Africa was the acknowledged leader in the field, maintaining a stable population in Kruger National Park while raising an average of US$1.4 million a year. In the case of Zimbabwe, revenues from culling and sport hunting were used not only to help maintain national parks but to provide compensation to rural peasants outside the park for losses caused by elephant crop-raiding, thereby encouraging them to regard elephants as a resource rather than a menace and to resist opportunities for poaching.

The position of both South Africa and Zimbabwe was that the proceeds of wildlife farming could be used to raise the standard of living for growing human populations without endangering the survival of wildlife, as their own experience showed. All this provided a significant underpinning for the policy of sustainable utilisation advocated by the IUCN, WWF and CITES. They argued in favour of cooperation with the ivory trade as the best means to control it.

Frustrated by the failure of the conservation establishment to take action against the trade, a group of Kenya-based biologists—Cynthia Moss, Joyce Poole, Iain Douglas-Hamilton and his wife Oria—decided to campaign for a ban on the trade. Their task was daunting. To get a ban imposed they needed to persuade the members of CITES that the African elephant was in danger of becoming extinct unless it was given Appendix I protection. Yet not even the WWF, renowned for its campaigns to save pandas and whales, favoured restrictions on the trade, let alone any government.

The strategy they adopted was to mobilise public opinion against the trade; and the target they chose was the CITES

conference scheduled for October 1989. Moss and Poole
used every opportunity to canvass the media, showing their
work in Amboseli to film crews and journalists from
around the world. Newspapers and magazines began taking
an increasing interest, investigating poaching and smuggling rackets and publishing photographs of the consequences: carcasses with hacked-off faces. In Washington,
Moss and Poole gained the support of the African Wildlife
Foundation, one of the main supporters of Amboseli's
research work, which agreed to sponsor an advertising
campaign. The advertising campaign was emotional and
hard-hitting. 'Today in America,' ran one advertisement,
'someone will slaughter an elephant for a bracelet.' Women
wearing ivory jewellery were labelled as 'Accessories to
Murder'. Other animal welfare organisations joined the
fray. Bumper stickers were produced, saying: 'Only Elephants Should Wear Ivory' and 'Save an Elephant, Shoot a
Poacher'. The African Wildlife Foundation named 1989 as
'The Year of the Elephant' and chose Cynthia Moss's book
*Elephant Memories*, with its tales of Echo, Teresia, and Slit
Ear, to launch it. Supporters in the US Congress passed the
African Elephant Conservation Act which authorised a ban
on ivory imports from states dealing in illegal ivory traffic.
Imports from Somalia, Gabon, Chad and Ethiopia were
immediately prohibited.

Similar campaigns were launched in Europe. In France,
the film actress Brigitte Bardot became the figurehead of an
'Elephant Amnesty' launched by the French scientist Pierre
Pfeffer. In Britain, the London *Daily Mail* started a 'Save
the Elephant' campaign. New organisations such as
Elefriends and Tuskforce sprang into life. The campaigns
were intense, using slogans, soundbites and gruesome

photographs to telling effect. At a time of mounting concern about issues such as global warming and diminishing rainforests, suddenly the African elephant became a potent symbol of the world's endangered environment.

Seeing the need for an independent body of specialists to ascertain the impact of the trade on elephants before the CITES meeting in 1989, Jonah Western organised funds for the establishment of an Ivory Trade Review Group. Set up in 1988, the ITRG consisted of a team of forty biologists, ecologists, economists, population modellers, trade specialists and investigators, coordinated by an Oxford zoologist, Steve Cobb. Its remit was to provide hard evidence for CITES delegates to enable them to take an informed view.

The CITES Secretariat decided to sponsor a rival trade study. With funds obtained from the Kowloon and Hong Kong Ivory Manufacturers Association, it recruited Ian Parker for the task. Parker, in turn, commissioned Graeme Caughley, an Australian population ecologist who had worked in Zambia's Luangwa Valley. Caughley's preliminary findings were grim. Ivory yields were in precipitous decline, he reported; exports were falling; the trade showed every sign of classic over-harvesting. If the trend continued, elephant populations would be 'commercially extinct' in East Africa in five years and over Africa as a whole within twenty years. Caughley's final report to CITES was never published.

Reports from the field, meanwhile, caused further alarm. There was mounting evidence that South Africa was acting as an entrepôt for illegal ivory smuggled in from other southern African countries. A US congressional committee was told in 1988 that anti-government rebels in Angola led by Jonas Savimbi were using ivory to pay for weapons supplied by the South African military. The South African military denied any responsibility. But Savimbi himself admitted that the South Africans had required him to pay in kind for supplies he received, some of it in ivory. Subsequently, a former South African special forces commander, Colonel Jan Breytenbach, who had been involved in Angola, confirmed to a Johannesburg newspaper that the rebels had mown down elephants indiscriminately. 'They shot everything,' Breytenbach remarked, 'bulls, cows and calves, showing no mercy in a campaign of extermination never before seen in Africa. The hundreds of thousands of elephants became thousands; the thousands became hundreds; and the hundreds, tens.' Similar activity was reported by anti-government rebels in Mozambique whom the South African military also supported. Trade statistics indicated South Africa was heavily engaged in ivory dealing, much of it illegal. From its own elephant culls, South Africa obtained no more than seven tons of ivory annually, but annual exports of ivory from South Africa in the mid-1980s reached nearly fifty tons.

The news from Kenya was equally bleak. The first full elephant count in Tsavo for eight years recorded in 1988 that the elephant population had halved. Only 5,300 elephants were found alive. Groups of orphans were a common sight; few males over the age of thirty-five were seen. Park rangers were known to be working in collaboration with gangs of poachers. In a number of incidents, they had shot elephants from park vehicles donated by conservation organisations for their protection. The Wildlife Department was riddled with corruption. When government ministers were provided with evidence of corrupt officials, they took no action.

Poaching gangs became ever bolder. Not far from the entrance to Tsavo park, a gang attacked a group of German tourists, shooting out the tyres of their vehicle and demanding money and other valuables. In the Kora National Park three rangers were shot dead. Not just elephants were under threat but Kenya's reputation as a leading tourist destination, with potentially severe repercussions. For tourism was Kenya's biggest source of foreign exchange, earning about US$500 million annually. When poachers killed Kenya's last surviving white rhinos in their enclosures in Meru National Park, President Moi finally stirred into action and appointed Richard Leakey, a distinguished palaeontologist and director of Kenya's National Museums, as head of a new Kenya Wildlife Service, with a mandate to root out corruption.

In May 1989, as public opinion in the West against the ivory trade gathered momentum, members of the Ivory Trade Review Group met in London to review their progress. It was a pivotal moment. At an earlier meeting, the evidence produced by population biologists had shaken even the most ardent advocates of a sustainable ivory trade. 'The results are a matter of biology, not economics,' remarked Professor David Pearce, head of the economics team from the London Environmental Economics Centre. The overall conclusion was that at existing levels of trade, the African elephant would be extinct within twenty-five to thirty years. Only by halting the trade now would there be any options left for the future. When members were asked to give their verdict, one by one they opted for placing the African elephant on Appendix I prohibiting international trade in ivory. Though the economists in the group doubted the effectiveness of a ban, they voted the same way.

Because of the danger that their decision would precipitate a sudden poaching upsurge and drive the trade underground in the four months before CITES met to consider an official ban, the ITRG recommended an immediate import ban in all consumer countries. Mindful of public opinion, one government after another duly announced a ban on imports. Britain and the United States were followed by Switzerland, Hong Kong, Canada, the European Community, Australia and, finally, Japan. With few buyers in the market, ivory prices plummeted.

African opinion turned too. Tanzania, after losing 200,000 elephants in ten years, two-thirds of its national herd, became the first African country to support an Appendix I listing. Kenya, under Richard Leakey's guidance, followed suit. In a dramatic publicity stunt, President Moi set fire to a stack of twelve tons of ivory piled thirty feet high in Nairobi National Park as a symbol of his new-found determination to crush the trade.

Southern African countries fought tenaciously to prevent a ban. Their herds were safe, they argued; they were expanding; they represented a valuable economic resource of benefit to rural communities. There was no reason why southern Africa should be penalised for the failure of other countries to protect their herds. Provided the trade was carefully monitored and policed, it should be kept open. If the East Africans needed lessons on wildlife management, they would be happy to oblige.

In an increasingly bitter war of words, the East Africans retorted that any continued trade would leave loopholes that traders were bound to exploit. Southern Africa's record in controlling illegal traffic, they argued, was not nearly as effective as its experts maintained. As long as a legal market remained, the whole of Africa's population was threatened.

The issue came to a head at the CITES conference in Lausanne, Switzerland in October 1989. After a stormy debate lasting two weeks, a compromise agreement was reached. The African elephant was transferred from Appendix II to Appendix I, thereby making it officially an endangered species; but special provisions were included to allow countries with healthy elephant populations and effective management programmes to downlist their populations.

As ivory prices continued to fall, the poaching epidemic subsided. But the cost had been high. During the 1980s Africa's elephant population had been halved. The remaining population numbered no more than 600,000. At least half a million had perished. Among the casualties were Douglas-Hamilton's Manyara elephants.

# 25
## An Image of Liberty

---

In his novel *The Roots of Heaven*, Romain Gary tells the story of a French prisoner-of-war held in a German concentration camp who survives by dreaming of elephants marching freely across the open plains of Africa. Released after the war he dedicates himself to their protection.

One can't spend one's life in Africa without acquiring something pretty close to a great affection for the elephants. Those great herds are, after all, the last image of liberty left among us. It's something that's fast disappearing, from more points of view than one. Every time you come upon them in the open, moving their trunks and their great ears, an irresistible smile rises to your lips. I defy anyone to look upon elephants without a sense of wonder. Their enormous size, their clumsiness, their gigantic stature represent a mass of liberty that sets you dreaming.

Many of the great herds have now gone. During the twentieth century, large parts of Africa lost their elephants irrevocably. Whereas forty-six countries once possessed elephant herds, now only thirty-five do. In twenty of those, elephant populations are largely insignificant. The sum total in fourteen countries in West Africa is no more than 18,000. Only five countries—Botswana, Zimbabwe, Tanzania, Congo-Kinshasa and Gabon—possess populations greater than 50,000 elephants.

Villagers in West Africa still tell stories of elephants and perform elephant masquerades, but few have ever seen one. For the vast urban populations of Africa, they are even more remote.

The future of many herds that have survived remains uncertain. The ban on international trade in ivory has provided some respite. Though poaching gangs are still active, they operate at a far lower level than before the ban. With ivory prices depressed, the rewards of the poaching business are no longer so attractive.

But other hazards that Africa's elephants had to endure during the twentieth century are just as threatening as before. Each year, human expansion into elephant areas reduces what territory is left to them to survive. The outcome, outside the boundaries of national parks, seems inevitable. Elephants have only a precarious future there.

In the long term, the survival of the remaining herds depends on the effective management of national parks. Extending for some 250,000 square miles, they provide elephants with their last refuge. But they are expensive to protect and vulnerable to attack from poaching gangs. Impoverished governments in Africa have few resources and many other priorities. Even though elephants are a star attraction for the tourist industry, past experience suggests this is no guarantee of their safety. Their future is further undermined by the proliferation of civil strife, lawlessness and corruption that afflicts many parts of Africa. After years of conflict in countries such as Angola, Congo-Kinshasa, Sudan and Somalia, no one has any clear idea of the fate of elephant populations there.

The role of the ivory trade, meanwhile, remains unresolved. Each year, as ivory stockpiles mount, the countries of southern Africa campaign for an end to the ban on international trade, arguing that it deprives them of revenues needed to protect their national parks and wildlife reserves; Zimbabwe, with an elephant population of more than 80,000, leads the field. Other African countries resist the move, claiming that any relaxation of the ban would drive up the price of ivory and encourage poaching.

In 1997 CITES agreed to allow Botswana, Namibia and Zimbabwe to make a one-time sale of sixty tons of stockpiled ivory to Japan. In 2000 South Africa asked CITES for approval to sell thirty tons on an 'experimental' basis, while Botswana, Namibia and Zimbabwe requested annual quotas. In the face of determined opposition, they deferred their demands. In November 2002, however, CITES voted by a narrow margin to allow Botswana, Namibia and South Africa to make another one-time sale of 60 tons of ivory from their stockpiles in 2004 provided that a new system for monitoring elephant poaching—called Monitoring of the Illegal Killing of Elephants or MIKE— had been properly established by then. Opponents of the sales doubted the new system would be effective enough to prevent an upsurge in poaching.

The clamour for approval to sell ivory stockpiles comes not only from southern Africa. Each year, most countries in eastern and southern Africa add at least one ton and in some cases five tons to their stockpiles; Zimbabwe accumulates about ten tons each year. As pressure grows for a change to the rules, the ban, in its present form, is unlikely to remain in place.

The one factor that has moved decisively in the elephant's favour in recent years is the work of field biologists. Their discoveries have revealed an ever more remarkable and complex animal. While elephants have always been held in high esteem since early civilisation, the new dimensions of elephant life uncovered by biologists—their family loyalties, their intricate social system, their powers of communication and their sense of compassion—have enhanced both public affection for them and recognition of their importance. With each passing year, their reputation for being among the most intelligent and sociable of our companions on earth has grown, and, with it, our understanding that as their world is diminished, so is our own.

# Select Bibliography

Adamson, George, *Bwana Game*, London, 1968

Aelian, *On the Characteristics of Animals*, Harvard University Press, 1958

Alexander, Shana, *The Astonishing Elephant*, New York, 2000

Alpers, Edward A., *Ivory and Slaves in East Central Africa: Changing Patterns of International Trade to the Late Nineteenth Century*, London, 1975

Aristotle, *History of Animals*, London, 1862

Arrian, *History of Alexander's Expedition*, London, 1729

Baker, Samuel W., *The Albert Nyanza, Great Basin of the Nile, and Exploration of the Nile Sources*, 2 vols, London, 1866

———, *The Nile Tributaries of Abyssinia*, London, 1867

———, *Wild Beasts and their Ways: Reminiscences of Europe, Asia, Africa and America*, London, 1890

Baldwin, William C., *African Hunting and Adventure*, London, 1864

Balfour, Daryl and Sharna, *African Elephant: A Celebration of Majesty*, Cape Town, 1997

Barbier, E.B., Burgess, J.C., Swanson, T.M., Pearce, D.W., *Elephants, Economics and Ivory*, London, 1990

Barker, W., 'The Elephant in The Sudan', in Hill (ed.), 1953

Beachey, R.W., 'The East African Ivory Trade in the Nineteenth Century', *Journal of African History*, 8, 1967

Bell, Walter D.M., *The Wanderings of an Elephant Hunter*, London, 1923, 1958

Bell, Walter D.M., *Bell of Africa*, London, 1960

Bere, Rennie, *The African Elephant*, London, 1966

Blunt, David E., *Elephant*, London, 1933

Bonner, Raymond, *At the Hand of Man: Peril and Hope for Africa's Wildlife*, London, 1993

Bosman, Paul, and Hall-Martin, Anthony, *Elephants of Africa*, Cape Town, 1986

Bosman, Willem, *A New and Accurate Description of the Coast of Guinea, divided into the Gold, the Slave and the Ivory Coasts*, London, 1705

Breasted, J.H., *Ancient Records of Egypt*, Chicago, 1906–7

Bruce, James, *Travels to Discover the Source of the Nile in the Years 1768–1773*, Edinburgh, 1790

Bryden, H.A., 'The Decline and Fall of the South African Elephant', *Fortnightly Review*, 79, London, 1903

Bull, Bartle, *Safari: A Chronicle of Adventure*, London, 1988

Burstein, Stanley M., *Agatharchides of Cnidus, On the Erythraean Sea*, London, 1989

Burton, Richard F., *The Lake Regions of Central Africa*, London, 1860

———, *Zanzibar: City, Island and Coast*, London, 1872

Buss, Irven, *Elephant Life: Fifteen Years of High Population Density*, Iowa State University Press, 1990

Cameron, Verney L., *Across Africa*, London, 1877

Carrington, Richard, *Elephants*, London, 1958

Chadwick, Douglas, *The Fate of the Elephant*, London, 1992

Conrad, Joseph, *The Heart of Darkness*, London, 1902

Cumming, R. Gordon, *Five Years of a Hunter's Life in the Far Interior of South Africa*, London, 1850

Davidson, Basil, *The African Past*, London, 1964

de Beer, Gavin, *Alps and Elephants: Hannibal's March*, London, 1955 de Brunhoff, Jean, *Histoire de Babar*, Paris, 1931; *The Story of Babar*, New York, 1933

De Watteville, Vivienne, *Speak to the Earth: Wanderings and Reflections among Elephants and Mountains*, London, 1936

Denis, Armand, *On Safari*, London, 1963

DiSilvestro, Roger L., *The African Elephant*, New York, 1991

Douglas-Hamilton, Iain, On the Ecology and Behaviour of the African Elephant, D.Phil.Thesis, University of Oxford, 1972

————, The African Elephant Action Plan, IUCN/WWF/NYZS Elephant Survey and Conservation Programme, 1979

Douglas-Hamilton, Iain and Oria, *Among the Elephants*, London, 1975

————, *Battle for the Elephants*, London, 1992

Drummond, W.H., *The Large Game and Natural History of South and South-East Africa*, Edinburgh, 1875

Eltringham, S.K., *Elephants*, Poole, 1982

Eltringham, S.K. (ed.), *The Illustrated Encyclopaedia of Elephants*, London, 1991

Farrant, Leda, *Tippu Tip and the East African Slave Trade*, London, 1975

Finaughty, William, *The Recollections of an Elephant Hunter*, Bulawayo, 1980

Freeman-Grenville, G.S.P., *The East African Coast. Selected Documents from the First to the Earlier Nineteenth Century*, Oxford, 1962

Gary, Romain, *The Roots of Heaven*, London, 1958

Gavron, Jeremy, *The Last Elephant: An African Quest*, London, 1993

Gordon, Nicholas, *Ivory Knights: Man, Magic and Elephants*, London, 1991

Graham, Alistair, *The Gardeners of Eden*, London, 1973

Gray, Richard, *A History of the Southern Sudan, 1839–1889*, London, 1961

Gray, Richard and Birmingham, David (eds.), *Pre-Colonial African Trade: Essays on Trade in Central and Eastern Africa before 1900*, London, 1970

Gröning, Karl and Saller, Martin, *Elephants: A Cultural and Natural History*, Cologne, 1998

Hall, Martin, *The Changing Past: Farmers, Kings and Traders in Southern Africa 200–1860*, Cape Town, 1987

Hall, Richard, *Empires of the Monsoon: A History of the Indian Ocean and its Invaders*, London, 1996

Hall-Martin, Anthony, *A Day in the Life of an African Elephant*, Halfway House, South Africa, 1993

Hanks, John, *A Struggle for Survival: The Elephant Problem*, Cape Town, 1979

Harms, Robert W., *River of Wealth, River of Sorrow: The Central Zaire Basin in the Era of the Slave and Ivory Trade, 1500–1891*, Yale University Press, 1981

Harris, William Cornwallis, *The Wild Sports of Southern Africa*, London, 1852

Hill, W.C.O. (ed.), *The Elephant in East Central Africa*, London, 1953

Hochschild, Adam, *King Leopold's Ghost*, London, 1999

Holder, Charles F., *The Ivory King: A Popular History of the Elephant and Its Allies*, London, 1886

Ivory Trade Review Group, The Ivory Trade and the Future of the African Elephant, Report prepared for the 7th Conference of Parties to CITES, Oxford, 1989

Jackson, Peter, *Endangered Species—Elephants*, London, 1990

Jeannin, Albert, *L'Éléphant d'Afrique: zoologie, histoire, folklore, chasse, protection*, Paris, 1947

Jolly, W.P., *Jumbo*, London, 1976

Select Bibliography

Kipling, Rudyard, *Just-So Stories*, London, 1986

Knight, Charles, *The Elephants Principally Viewed in Relation to Man*, London, 1844

Krapf, Johann L., *Travels, Researches and Missionary Labours During an Eighteen Years' Residence in Eastern Africa*, London, 1860

Künkel, Reinhard, *Elephants*, New York, 1982

Laws, R.M., Parker, I.S.C., and Johnstone, R.C.B., *Elephants and their Habitat: The Ecology of Elephants in North Bunyoro, Uganda*, Oxford, 1975

Livingstone, David, *Missionary Travels and Researches in South Africa*, London, 1857

————, *Narrative of an Expedition to the Zambezi and its Tributaries*, London, 1865

————, *Last Journals*, London, 1874

Livy, Titus, *The War with Hannibal,* London, 1972

Ludolphus, The Learned Job, *A New History of Ethiopia being a Full and Accurate Description of the Kingdom of Abessinia vulgarly, Though erroneously called the Empire of Prester John*, London, 1682

Masson, Jeffrey and McCarthy, Susan, *When Elephants Weep: The Emotional Lives of Animals*, London, 1994

Moore, E.D., *Ivory, Scourge of Africa*, New York, 1931

Moore, Randall J. and Munnion, Christopher, *Back to Africa*, Johannesburg, 1989

Moorehead, Alan, *The White Nile*, London, 1960

————, *The Blue Nile,* London, 1962

Moss, Cynthia, *Portraits in the Wild: Animal Behaviour in East Africa*, London, 1989

————, *Echo of the Elephants*, London, 1992

————, *Little Big Ears: The Story of Ely*, New York, 1997

————, *Elephant Memories*, University of Chicago, 2000

Neumann, Arthur H., *Elephant-Hunting in East Equatorial Africa, being an account of three years ivory-hunting under Mount Kenia and among the Ndorobo savages of the Lorogi mountains, including a trip to the north end of Lake Rudolf*, London, 1868

Offermann, P.P.M., 'The Elephant in the Belgian Congo', in Hill (ed.), 1953

Orenstein, Ronald, *Saving the Gentle Giants*, London, 1991

Osborn, Henry F., *Proboscidea: A Monograph on the Discovery, Evolution, Migration and Extinction of the Mastodonts and Elephants of the World*, 2 vols, New York, 1936, 1942

Oswell, W.E., *William Cotton Oswell*, London, 1900

Park, Mungo, *Travels in the Interior Districts of Africa: Performed under the Direction and Patronage of the African Association in the Years 1795, 1796 and 1797*, London, 1799

Parker, I.S.C., The Ivory Trade, 3 vols, Report prepared for US Fish and Wildlife Service and IUCN, Nairobi, 1979

Parker, Ian and Amin, Mohamed, *Ivory Crisis*, London, 1983

Payne, Katy, *Silent Thunder: The Hidden Voice of Elephants*, London, 1998

Perry, John, 'The Growth and Reproduction of Elephants in Uganda', *The Uganda Journal*, 16 (1), 1952

Petherwick, Mr and Mrs John, *Travels in Central Africa*, 2 vols, London, 1869

Pfeffer, Pierre, *Vie et Mort D'Un Géant: L'Éléphant D'Afrique*, Paris, 1989

Pinkerton, John, *A General Collection of the Best and Most Interesting Voyages and Travels in all Parts of the World: Many of Which are now Translated into English*, London, 1814

Pitman, Charles, 'The Elephant in Uganda', in Hill (ed.), 1953

Pliny the Elder, *Natural History*, 2 vols, London, 1855

Polybius, *The General History of Polybius*, Oxford, 1823

Poole, Joyce, *Coming of Age with Elephants*, London, 1996

Ranking, J., *Historical Researches on the Wars and Sports of the moguls and Romans: in which Elephants and Wild Beasts were employed or slain*, London, 1826

Rodney, Walter, *A History of the Upper Guinea Coast, 1545–1800*, Oxford, 1970

Roosevelt, Theodore, *African Game Trails*, New York, 1910

Ross, Doran H. (ed.), *Elephant: The Animal and Its Ivory in African Culture*, University of California, 1992

Rushby G.G., 'The Elephant in Tanganyika', in Hill (ed.), 1953

———, *No More The Tusker*, London, 1965

Russell, Peter, *Prince Henry, 'the Navigator'*, Yale University Press, 2000

St Aubyn, Fiona, *Ivory: A History and Collector's Guide*, London, 1987

Sanderson, Ivan T., *The Dynasty of Abu: A History and Natural History of the Elephants and their Relatives Past and Present*, London, 1960

Schweinfurth, Georg, *The Heart of Africa: Three Years' Travels and Adventures in the Unexplored Region of Central Africa from 1868 to 1871*, London, 1873

Scullard, H.H., *The Elephant in the Greek and Roman World*, London, 1974

Selous, Frederick, *A Hunter's Wanderings in Africa*, London, 1881

———, *Travel and Adventure in South-East Africa*, London, 1893

Sheriff, Abdul, *Slaves, Spices and Ivory in Zanzibar*, London, 1987

Shoshani, Jeheskel (ed.), *Elephants: Majestic Creatures of the*

*Wild*, San Francisco, 1992

Sikes, Sylvia K., *The Natural History of the African Elephant*, London, 1971

Sillar, F.C. and Meyler, R.M., *Elephants Ancient and Modern*, London, 1968

Sparrman, Anders, *A Voyage to the Cape of Good Hope towards the Antarctic Polar Circle, and Round the World: but Chiefly into the Country of the Hottentots and Caffres, from the Year 1772 to 1776*, 2 vols, London, 1786

Spinage, Clive, *Elephants*, London, 1994

Stanley, Henry M., *How I Found Livingstone: Travels, Adventures and Discoveries in Central Africa*, London, 1872

———, *Through the Dark Continent*, 2 vols, London, 1878

———, *The Congo and the Founding of the Free State*, London, 1885

———, *In Darkest Africa: or, The Quest, Rescue and Retreat of Emin, Governor of Equatoria*, 2 vols, London, 1890

Stevenson-Hamilton, James, *Wild Life in South Africa*, London, 1947

Stigand, C.H., *Hunting the Elephant in Africa, and other Recollections of Thirteen Years' Wanderings*, London, 1913

Stockley, Charles, 'The Elephant in Kenya', in Hill (ed.), 1953

Swann, Alfred J., *Fighting the Slave-Hunters in Central Africa*, London, 1910

Temporal, Jean-Luc, *La Chasse Oubliée*, Paris, 1989

Thomson, Joseph, *To the Central African Lakes and Back*, London, 1881

Thorbahn, Peter F., The Pre-Colonial Ivory Trade of East Africa, PhD Thesis, University of Massachusetts, 1979

Thornton, Allan and Currey, Dave, *To Save an Elephant: The Undercover Investigation into the Illegal Ivory Trade*,

London, 1991

Tippu Tib, *Maisha ya Hamed bin Muhammed el Murjebi, Yaani Tippu Tib*, Nairobi, 1966

Topsell, Edward, *The Historie of the Foure-Footed Beasts: describing the true and lively figure of every Beast . . . collected out of all the volumes of Conradius Gesner; and all other writers of this present day*, London, 1607

Toynbee, J.M.C., *Animals in Roman Life and Art*, London, 1973

Turner, Myles, *My Serengeti Years: The Memoirs of an African Game Warden*, London, 1987

Weber, Nicholas F., *L'Art de Babar: L'oeuvre de Jean et Laurent de Brunhoff*, Paris, 1989

Western, David, *In the Dust of Kilimanjaro*, Washington, 1997

Williams, Heathcote, *Sacred Elephant*, London, 1989

Wilson, Derek and Ayerst, Peter, *White Gold: The Story of African Ivory*, London, 1976

# Index

# Index

# Index

# Index

# Picture Acknowledgements

Illustrations in the text: Thomas Bewick, *General History of Quadrupeds*, 1785. The British Library. G. Cuperus, *De Elephantis in nummis obviis exercitationes duae*, 1735. Mary Evans Picture Library, London. Conrad Gesner, *Historiae Animalium*, Zürich, 1551-87. Captain William Cornwallis Harris, *The Wild Sports of Southern Africa*, London, 1844. Charles Frederick Holder, *The Ivory King*, London, 1886. Hulton Archive, London. *The Illustrated London News*. A. Jeannin, *L'Eléphant d'Afrique*, Paris, 1947. Rudyard Kipling, *Just So Stories for Little Children*, London, 1902, Macmillan & Co. Ltd. Private Collections. Frederick Courteney Selous, *A Hunter's Wanderings in Africa*, London, 1895. Stapleton Collection/Bridgeman Art Library, London. P.L. Shinnie, *Meroe*, 1967. Sir J. Emerson Tennent, *The Wild Elephant*, London, 1867. Edward Topsell, *The Historie of Foure-Footed Beasts*, London, 1607.

Colour photographs: ©Lee Lyon by permission Iain and Oria Douglas-Hamilton, Save the Elephants: page 1. HPH Photography/Bruce Coleman Collection: page 2 top. ©Steve Bloom Images: page 2 bottom, page 4. BBC Natural History Unit Picture Library: ©Anup Shah page 3 top, ©Pete Oxford page 6, ©Richard Du Toit page 7 top. ©Joyce Poole: page 3 bottom, page 5 top. Bruce Coleman Collection: page 5

*Picture Acknowledgements*

centre. ©Wendy Dennis/FLPA Frank Lane Picture Agency: page 5 bottom. ©Cynthia Moss page 7 bottom. Art Wolfe/Science Photo Library: page 8.

PUBLICAFFAIRS is a publishing house founded in 1997. It is a tribute to the standards, values, and flair of three persons who have served as mentors to countless reporters, writers, editors, and book people of all kinds, including me.

I. F. STONE, proprietor of *I. F. Stone's Weekly*, combined a commitment to the First Amendment with entrepreneurial zeal and reporting skill and became one of the great independent journalists in American history. At the age of eighty, Izzy published *The Trial of Socrates*, which was a national best-seller. He wrote the book after he taught himself ancient Greek.

BENJAMIN C. BRADLEE was for nearly thirty years the charismatic editorial leader of *The Washington Post*. It was Ben who gave the *Post* the range and courage to pursue such historic issues as Watergate. He supported his reporters with a tenacity that made them fearless, and it is no accident that so many became authors of influential, best-selling books.

ROBERT L. BERNSTEIN, the chief executive of Random House for more than a quarter century, guided one of the nation's premier publishing houses. Bob was personally responsible for many books of political dissent and argument that challenged tyranny around the globe. He is also the founder and was the longtime chair of Human Rights Watch, one of the most respected human rights organizations in the world.

· · ·

For fifty years, the banner of Public Affairs Press was carried by its owner Morris B. Schnapper, who published Gandhi, Nasser, Toynbee, Truman, and about 1,500 other authors. In 1983 Schnapper was described by *The Washington Post* as "a redoubtable gadfly." His legacy will endure in the books to come.

Peter Osnos, *Publisher*